SEED PRODUCTION TECHNIQUES OF MAJOR CROPS

ABOUT THE AUTHORS

Dr. O.P. Verma did B.Sc. (Ag.) Hons. from U.P. Institute of Agricultural Sciences, Kanpur. Later he completed his M.Sc. (Ag.) and Ph.D. in Plant Breeding and Genetics from N.D. University of Agriculture and Technology, Kumarganj, Faizabad and Kanpur University, Kanpur, respectively. He is teaching the courses on Seed Technology and Plant Breeding at under graduate and Post graduate levels since 1985 and guided M.Sc. (Ag.) and Ph.D. students. He has published more than 60 research papers in National and International Journals and also contributed in the development of medium duration high yielding variety of rice (Narendra–359) release from CVRC.

Presently he is working as Breeder (R&M)/Asso. Professor in the Department of Genetics and Plant Breeding. He is born in Agriculture family at Lakhimpur (Kheri) of Uttar Pradesh. He has rich experience of Plant Breeding in general and Seed Technology in particular.

Dr. A.L. Jatav born in 1962 at district Bareilly, U.P. did B.Sc. (Ag. & AH) from GBPUALT Pantnagar, M.Sc. in Seed Science & Technology and Ph.D. in Mycology and Plant Pathology from IARI, New Delhi. He started his carrer as Asstt. Professor in the Deptt. of Seed Science & Technology CSAA A&T, Kanpur are remained these till 2009. In the meantime he worked as Asstt. Commissioner, Seed Quality Control, Seed Division of DAC, Govt. of India during 2000-01.

At present he is working as Professor Life Science Deptt., CSJM University, Kanpur. He has guided several, M.Sc. & Ph.D. students and published number of research papers.

SEED PRODUCTION TECHNIQUES OF MAJOR CROPS

By

O.P. VERMA

2015
Daya Publishing House®
A Division of
Astral International Pvt. Ltd.
Delhi - 110 002

Published by   : **Daya Publishing House®**
                       A Division of
                       Astral International Pvt. Ltd.
                       – ISO 9001:2008 Certified Company –
                       4760-61/23, Ansari Road, Darya Ganj
                       New Delhi-110 002
                       Ph. 011-43549197, 23278134
                       E-mail: info@astralint.com
                       Website: www.astralint.com

Laser Typesetting : **Classic Computer Services**
                       Delhi - 110 035
Printed at   : **Chawla Offset Printers**
                       Delhi - 110 052

PRINTED IN INDIA

PREFACE

History continues to be a story of a "hungary man in search of food." Our life and health is dependent on seed and their products. With the passage of time, the problem of mankind are becoming more complicated because of ever increasing population and decreasing availability of food, space and money. The increasing pressure of population will demand more and more food. This problem calls for vertically increasing food productivity by ensuring availability of quality seeds of high yielding varieties and better management to achieve food security.

It is anticipated that by the year 2020, total food grains demand will reach 324 million tonnes against the supply of 314 million tonnes leaving a deficit of 10 million tonnes. Production of new varieties and hybrids in adequate quantities of high purity and quality largely depends upon trained human resources in this field who are conversant with all aspects of seed production.

In this book attempt has been made to bring the desired informations on quality seed production of different crops at one place for which I acknowledge, respectfully to the authors of different books which provided information and base to this attempt in valuable manner. We very sincerely hope that the information compiled in this book shall be extremely useful to the students of Seed Technology and persons engaged in seed production.

O.P. Verma

CONTENTS

1

PLANT VARIETY TESTING, RELEASE, NOTIFICATION AND MAINTENANCE

The activity of a plant breeder terminates with the release and notification of the variety and from here onward the Seed Technologist takes over, therefore, it is important for the Seed Technologist to know the system of variety release and notification in the country.

The ultimate goal of crops improvement programmes is the development and release of a higher and stable performing cultivar. The adoptability of a particular ecological situation is determined by testing a genotype over several locations for a few years. The unique advantage of multilocation testing that it provides in a few seasons the data equivalent to multi-seasons testing at a single location. In many crops species, both autogamous and allogamous, the cereals like wheat and rice, pulses and oilseeds, sorghum, pearlmillet and sunflower etc. several hundred of lines/crosses are being evaluated in multilocation trails for evaluation of a number of superior genotypes for different agro-ecological situations under the crop specific, All India Co-ordinated Projects.

The Co-ordinated projects with the objectives of evaluating varieties and other production technologies organise variety evaluation over locations and seasons through a network of research centres operating in SAU's, Central Institutes, ICAR Institutes and State Department of Agriculture. The general out line of a variety testing system is more or less similar irrespective of the breeding system *i.e.,* self or cross pollinated (Fig. 1-4).

```
┌──────────────┐                          ┌──────────────┐
│International  │                          │Station Trials│
│Nurseries     │                          └──────────────┘
└──────────────┘        ╭─────────╮
┌──────────────┐        │Promoting │       ┌──────────────┐
│State Multipli-│       │entries   │       │Disease/pest  │
│cation Trials │        │Identified at│     │Screening     │
└──────────────┘        │Research  │       │Nurseries     │
┌──────────────┐        │Centres   │       └──────────────┘
│Germplasm     │        ╰─────────╯
│evaluation    │
│Nurseries     │
└──────────────┘
```

	Initial Varietal/Hybrid Trials (IVT) (IHT)
Multilocational Ist Year	Advance Varietal/Hybrid Trials (AVT) (AHT)
Multilocational 2nd Year	Variety Identification
	Tested Stock Seed Multiplication
	Mini kits, On-Farm Trials
	Variety Release

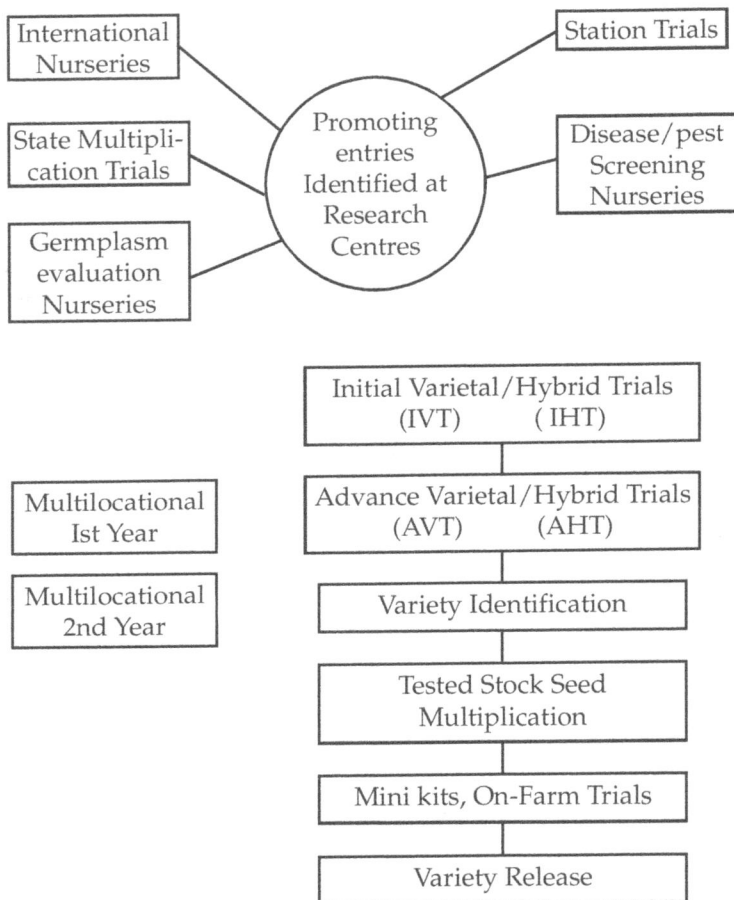

Fig. 1 : Variety Testing – General.

The multilocation testing that takes into account diverse agro-ecological needs, multi-disciplinary approach for performance evaluation, well defined trial plans from layout to data processing at regional, zonal and national level, assessment of varietal performance during scientists meeting and identification of varieties, on-farm testing to assess cultivator's reactions are some of the unique features of the co-ordinated research programmes.

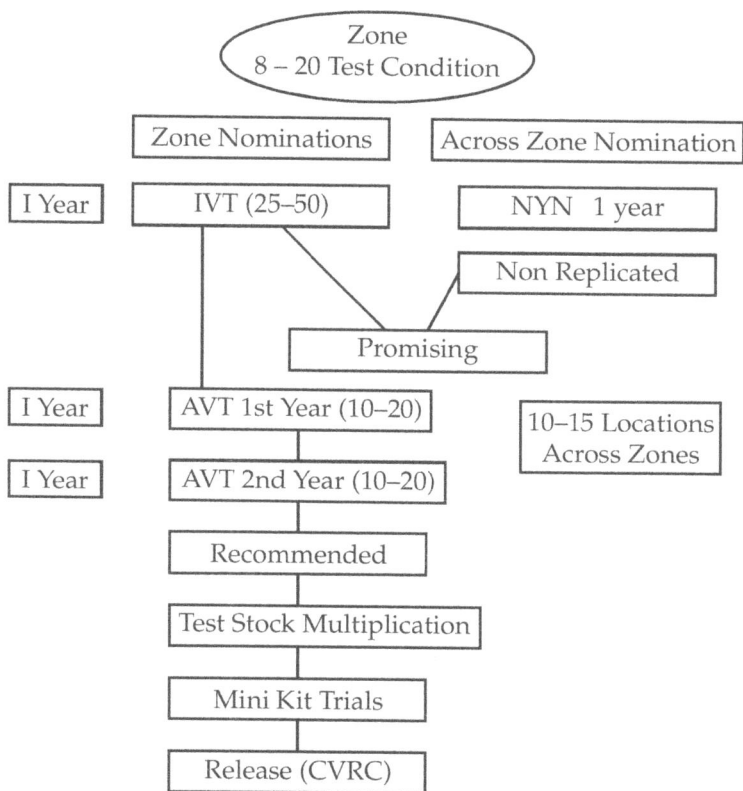

Fig. 2 : Three tier Testing – Wheat.

Testing Net Work

Before the entries included in the national variety testing system, undergo screening/evaluation at different stages in station trials conducted by breeders at the respective centres. At the time of annual meetings of the crop scientists, the test entries assembled at national or zonal levels are considered for evaluation uniformaly in a 3-tier system for a period of 3-4 years. Progressively the trials included less entries but are increasingly intensive in evaluation, the first stage of testing generally called IVYT/IET (Initial Variety Yield Trial/Initial Evaluation Trial) includes largest number of entries. Depending

```
                    ┌──────────────────────────┐
                    │   Zone Ist Year (NSP)     │
                    └──────────────────────────┘
┌──────────────────────┐              ┌──────────────────────────┐
│  Best from INGER etc. │              │ Across Zone best entries │
│ (Ecoys temperature wise)│            └──────────────────────────┘
└──────────────────────┘
┌──────────┐   ┌──────────────────────────────┐
│ 2 years  │   │   IVTC (Irrigated)/RF/Hills   │
└──────────┘   └──────────────────────────────┘
┌──────────┐   ┌──────────────────────────────┐
│ 2 years  │   │   AVTC (Irrigated)/RF/Hills   │
└──────────┘   └──────────────────────────────┘
               ┌──────────────────────────────┐
               │           Agronomy            │
               └──────────────────────────────┘
               ┌──────────────────────────────┐
               │     Variety identification    │
               └──────────────────────────────┘
               ┌──────────────────────────────┐
               │          Mini Kits            │
               └──────────────────────────────┘
               ┌──────────────────────────────┐
               │        Release (CVRC)         │
               └──────────────────────────────┘
```

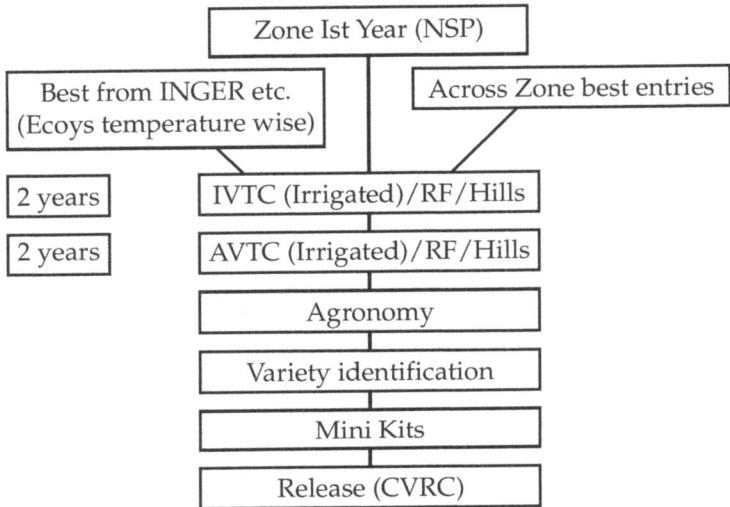

Fig. 3 : Three tier Testing – Rice.

on the diversity of the crop growing situations and stress environments, the genotypes are grouped duration wise and tested as replicated yield trial. Based on first year performance (10–15% more yield as compared to zonal check and national checks) the most promising 20–30% of the entries are promoted to the second stage of testing. Some times IVT/IET are conducted as unreplicated screening nursery in selected locations.

With the second stage of testing the number of entries is less but the plot size/ replications and locations are increased. The trials are constituted according to zone-wise or for the entire country or maturity duration which again depends on the crop. After two years of testing, the best performing (10–15%) are promoted to final stage of 3-tier testing. The final trial having a limited number of promising entries is conducted over a number of locations representing the ecosystem across the zone/country. The plot size is larger and the test entries are evaluated for yield and other ancillary traits in comparison to best local, regional and national checks for one or two years again depending on the crop. The best one/two/three entries are recommended for on-farm testing by the Ministry of Agriculture and subsequent

```
┌──────────────────────────────────────┐
│    State Plant Breeding Programme     │
└──────────────────────────────────────┘
                   │
┌──────────────────────────────────────┐
│   Multilocation Testing in the State  │
└──────────────────────────────────────┘
                   │
┌──────────────────────────────────────┐
│         All India testing under       │
│         Co-ordinated Projects         │
└──────────────────────────────────────┘
                   │
┌──────────────────────────────────────┐
│   Research Evaluation Committee       │
│         of the University             │
└──────────────────────────────────────┘
                   │
┌──────────────────────────────────────┐
│               Identified              │
└──────────────────────────────────────┘
                   │
┌──────────────────────────────────────┐
│            Adaptive Trials            │
└──────────────────────────────────────┘
                   │
┌──────────────────────────────────────┐
│                  RES                  │
└──────────────────────────────────────┘
                   │
┌──────────────────────────────────────┐
│   State Variety Release Committee     │
└──────────────────────────────────────┘
       │           │            │
┌────────────┐ ┌────────────┐ ┌──────────────┐
│    Seed    │ │ Package of │ │ Notification │
│Multiplication│ │  Practices │ │              │
└────────────┘ └────────────┘ └──────────────┘
```

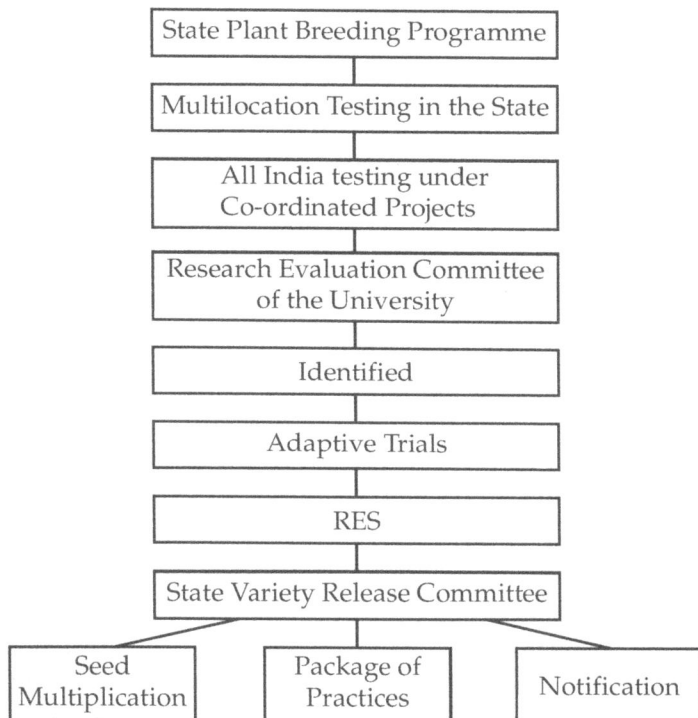

Fig. 4 : State Variety Testing and Release.

release of any one or all/none of them by the central sub committee on Crop Standard, Notification and Release of Varieties. This committee consists of Scientists, Directors of Agriculture and Seed Certification Agencies, Seed Companies and Progressive farmers. Deputy Director General (Crop Science) of ICAR is the chairman of the committee. If the proposed variety is found superior than the ruling variety, it is recommended to the central seed committee for its release and notification with minimum standard limits and it becomes eligible for certification.

States can release any of their entries under testing/tested in the coordinated programme at any stage based on their own state level yield/adaptive/on farm trial data through the State Seed Sub Committee.

Period of Testing

Period of testing varies from crop to crop. It is 5 years in rice (1 + 2 + 2), 4 years in wheat, 3 years (1 + 2 or 2 +1) in groundnut, sorghum, pulses, oilseeds including cross pollinated crops. Among all crops rice is grown under most diverse agro-ecologies. Similarly the pest disease syndrome has been changing much fast in rice in comparison to other crops. In vegetable crops, there is no programme like Initial Evaluation Trial (IET). Only Advance Varietal Trial (AVT) for 3 years are conducted.

Data Collection and Reporting

Uniform/Standardized data sheets have been developed over years. The data sheets are different for different crops. The data are generated on seed yield and on agronomic/economic importance. Reaction to diseases/pests are an integral part of the data sheet. For specific diseases hot spots/endemic areas have been identified to generate reliable data on disease and pest reaction.

Quality characters, depending on their relative importance are analysed at one or the other stage. In the final stage of testing only few selected entries are evaluated for agronomic/protection, characters *e.g.*, date of sowing, response to different levels of nutrients, plant, water management and other cultural practices in the agronomic trials.

Data Processing and Variety Identification

The data received from co-operating Centres are pooled and analysed as per the statistical design. Depending on the experimental design ANOVA is done to bring out varietal differences. Yield superiority in terms of over all mean, zonal yield and frequency of locations with top performance and the general criteria used for identifying the best varieties. While screening the data, the test locations with high CV (>30% for rainfed, >20% for irrigated) and location mean yield less than state average yield of that crop are excluded from analysis.

Besides data based assessment of the test entries, site visits by the monitoring teams comprising breeders of the region along with pathologists/entomologists/agronomists at appropriate growth stage help to visually identify the best performers. In addition the visit also helps in knowing the level of management of the trial and precision with which the trial is conducted.

The concerned breeders normally make a proposal for identification of their varieties based up on its performance/superiority. The criteria used by identification are –

❐ Significant superiority in yield over the best check.

❐ Yield superiority by 10% over the best check.

❐ Comparable in yield performance to the best check but possessing better quality and/or resistance to major diseases/insect pests in the given zone.

On the basis of the recommendations of the variety identification committee, the concerned breeders formally send release proposals of such identified entries to the central sub committee on Crop Standard, Notification and Release of Varieties. After release and notification, a variety qualifies for certification of its seed multiplication programmes.

Variety Maintenance

The availability of good quality seed to the farmers is the only way to realize the benefits of an improved variety. Maintaining a high quality nucleus seed (N/S) is broadly termed as maintenance breeding (MB). The dibbling of single seed by the breeder to initiate the seed production chain is the starting point of maintenance breeding.

After release of variety, it becomes necessary to maintain its genetic purity so that the originally of the variety is maintained not only for higher productivity but also for the identity of the variety and maintenance of quality of the produce. For achieving this objective information on diagnostic characteristics of the varieties/parental lines/hybrids is essential which can be used by the seed growers/seed corporations/seed certification agencies and also seed testing labs in order to remove off-types from the crop and to determine the genetic purity. The proper procedures for maintenance breeding *i.e.* nucleus and breeder seed production are also essential for maintenance breeding so maintain the originality of the variety.

In self pollinated crops :

❑ Select approx. 500 true to type ears/panicles (cereals), plants in other crops at maturity.

❑ Harvested and threshed separately. Examine the seeds for uniformity (colour, size, shape etc.). Reject plants showing variation.

❑ Grow (dibbling method) ear to row progenies or single plant progenies.

❑ Reject entire progeny showing mixture or non-uniformity even on the basis of a single off-type plant in the progeny.

The progeny of NSS–II is the breeder seed.

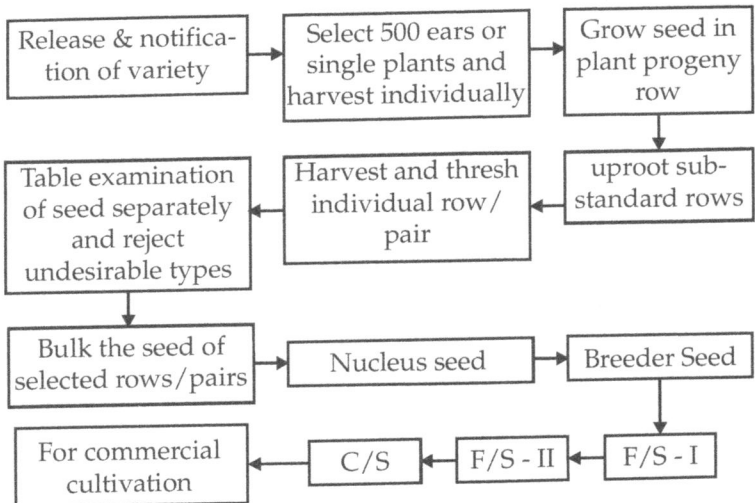

Release & notification of variety	→	Select 500 ears or single plants and harvest individually	→	Grow seed in plant progeny row

Table examination of seed separately and reject undesirable types	←	Harvest and thresh individual row/ pair	←	uproot sub-standard rows

Bulk the seed of selected rows/pairs	→	Nucleus seed	→	Breeder Seed

For commercial cultivation	←	C/S	←	F/S - II	←	F/S - I

```
              ┌─────────────────────┐
              │  Basic seed obtained │
              │ from evolving institute │
              └─────────────────────┘
```

Single plant produce		Bulk

Harvesting

```
  o  #  o  o  #
  o  o  #  o  o
  o  o  #  o  o
```

Use produce of marked single plant as source seed for production of Nucleus Seed	Use produce of unmarked plants as Nucleus seed

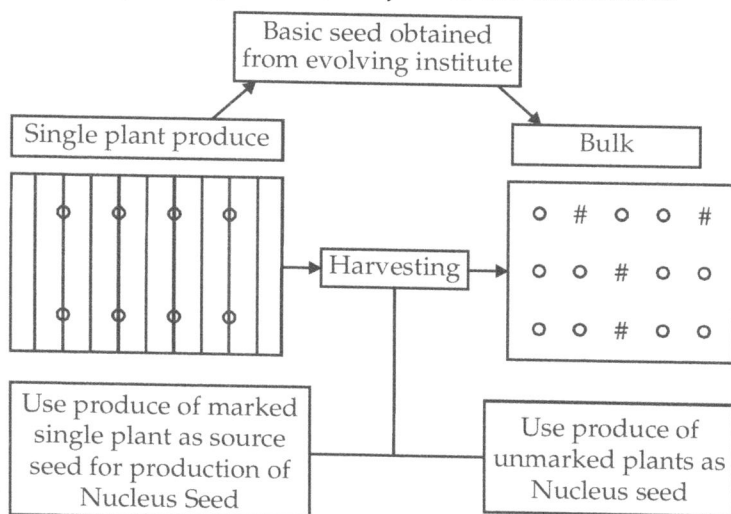

Production of Nucleus seed for the first time through original stock.

❑ Progeny rows true to type (phenotypically) are further selected, harvested separately and examined. Bulk the seed-called Nucleus seed stage–1 (NSS-I).

❑ Before bulking keep a portion of seed of each selected progeny separate for advancing to NSS–II.

MAINTENANCE OF N/S YEAR AFTER YEAR

I Year

❑ Required number of true to type plants/ear heads (wheat and rice) are selected at early stage and tagged in B/S plot.

❑ Tagged plants are repeatedly observed for distinguishing characters at different stages of crop growth.

❑ Plants exhibiting variation at any stage are rejected.

❑ Diseased plants are also rejected.

❏ Individual plant/ear head is properly kept in separate envelopes.

II Year

❏ Produce of selected plant/ear is sown in one row.

❏ Line to line and plant to plant provide more space as compared to commercial crop.

❏ Seeds of one envelope are never used for sowing of more than one row.

❏ Individual plant progenies are visited regularly to observe off types.

❏ The row is rejected in which off-type plants appear.

❏ Diseased plants are also rogued out.

❏ At maturity seed characters of individual plant are confirmed.

❏ The produce of selected lines is bulked and used as N/S for production of B/S.

In hybrid crops one has to deal with maintenance of male sterile line (A), maintainer (B) and restorer line (R). Parental lines are maintained separately in isolated plots by plant or ear to row method. N/S of female line (A) is maintained by undertaking mass selection in B-line. Ear to row progenies by B-line are grown in isolation adjacent to A-line. The off-type progenies of A and B-lines are removed. Seed parent rows are hand pollinated by collecting pollen from desirable B-line progenies. Individual plants of B-line are selfed. Similarly the parental stock of R-lines are maintained in isolation.

Infrastructure in India

The B/S production is the mandate of ICAR and is being undertaken with the help of :

❏ ICAR Research Institutes/National Research Centres/AICRPs.

❏ State Agricultural Universities.

❑ Sponsored B/S production through NSC (National Seeds Corporation) State Farms Corporation of India (SFCI) and Selected State Seeds Corporations*

❑ Non-governmental organisation (NGOs)*

Releasing the importance of seed in the country, the Govt. of India/ICAR has created enough infrastructure. There are 54 BSP units for different crops located in various SAU's and ICAR Institutes. In addition separate units for the production B/S of Groundnut, sunflower and soybean have been established at 24 centres.

Procedure for Production

Indents for B/S are placed by State Govt. to DAC, GOI who compiles and passes them to ICAR for excecution. At ICAR level, the production of B/S is allocated to different institutions at the Annual Workshop meeting of the respective crop. The final produced quantity of the B/S is reported back and allocated by the DAC to the indenters producing F/S.

Monitoring of Breeder Seed

The B/S is not certified however, for maintenance of quality the ICAR has constituted monitoring team for various crops. The monitoring system has helped to maintain the quality of B/S to a great extent. The Monitoring Team include!

❑ Crop Breeder/Representative of crop coordinator (ICAR).

❑ Producing Breeder.

❑ Representative of State Seed Certification Agency.

❑ Representative of National Seed Corporation.

Information Flow

To regulate and systematise the programme, various Breeder Seed Production (BSP) Proformance have been devised:

---

* Needed supervision by qualified breeder is provided to them to ensure the quality of seed produced.

BSPI – Provided by the Project Co-ordinator to ICAR, the Seed Division, GOI and the producing breeders indicating the allocation of production schedule, quantity and name of the members of monitoring team.

BSP II – Submitted by the producing breeder to ICAR Seed Division, DAC giving the details of sowing date, area, quantity targetted, expected date of monitoring.

BSP III – Report of monitoring team, which is submitted to ICAR and Seed Division, GOI after monitoring and contains information about genetic purity along with other recommendations.

BSP IV – This contains the final figures of the quantity of B/S produced (processed seed) and submitted by the concerned breeder to DAC/ICAR.

BSP V – It contains the final figures of the quantity of seed lifted/non-lifted as per allocation by the Seed Division, Ministry of Agriculture.

BSP VI – This proforma is filled by DAC and sent to ICAR and contains the details of quantity of B/S allotted/lifted and the F/S produced from the lifted one.

APPENDIX – I

Proforma for submission of proposal for identification of crop varieties/hybrids by the All-India Workshop.

1. (a) Name of the variety/hybrid :　　.............................

   (b) Species :　　.............................

2. Parentage :　　.............................

3. Breeding method used :　　.............................

4. Developed by (Station and Name of worker) :

   ....

5. Proposed by :　　.............................

6. Zone for which to be identified :

   ....

7. Production condition for which to be identified :

   ....

8. Incase of hybrid description of parents :

   ....

9. List atleast two important morphological features of the proposed variety/hybrid which distinguish it from other important commercial varieties under fieled conditions:

   ....

10. The new variety/hybrid provides an alternative replacement for :　　.............................

11. List main problems and special requirement (in order of importance) of the concerned area of recommendation and how the proposed variety helps to resolve these problems :　　.............................

12. Year when first entered in Coordinated Varietal Trials :
    ....

13. Quality of B/S available (*a*) variety :..............................

    (*b*) Parental lines in case of hybrids :............................

14. Summary data of coordinated varietal trials 3 years :
    ....

15. Problems and prospects in seed production of parental lines and hybrids and their maintenance (where-ever applicable) :   ....

APPENDIX – II

Proforma for Submission of proposal for release of Crop variety to Central Sub-Committee on Crop Standard, Notification and Release of Varieties.

1. Name of the Crop species   ....

2. (*a*) Name of the variety under which tested : ....

   (*b*) Proposed name of the variety :   ....

3. Sponsored by :   ....

4. (*a*) Institution or agency responsible for the development of the variety : ....

   (*b*) Name of the persons who helped in the development of the variety : ....

5. (*a*) Parentage with details of its pedigree : ....

   (*b*) Source of material in case of introduction :

        ....

   (*c*) Breeding method :   ....

   (*d*) Breeding objective :   ....

6. State the variety, which most closely resembles the proposed variety in general characteristics :

        ....

7. (*a*) Whether recommended by seminar/conference/workshop/State Seed Committee (SVRC) :

        ....

   (*b*) If so its recommendation with specific justification for the release of proposed variety : ....

(c) Specific areas of its adaptations : ....

8. Recommended ecology :   ....

9. Description of variety/ Hybrid

(a) Plant height :   ....

(b) Distinguishing morphological characteristics :

....

(c) Maturity group :   ....

(d) Reaction of major diseases (under field and controlled conditions) :   ....

(e) Reaction to major pests (under field and controlled conditions including store pests) : ....

(f) Agronomic features (*e.g.*, Resistance to lodging, shattering, fertilizer responsiveness, suitability for early or late sown conditions, seed rate etc.) :

....

(g) Quality of produce of grain, forage/fibre including nutritive value where-ever relevant:

....

(h) Reaction to stresses :   ....

10. In case of hybrid, description of parents :

....

11. (a) Yield data in regional/inter regional/district trial year wise (level of fertilizer application, density of plant population, and superiority over local/ standard varieties to be indicated) : ....

(b) Yield data from national demonstrations/large scale demonstrations :   ....

(c) Average yield under normal conditions : ....

12. (a) Agency responsible for maintaining B/S : ....

(b) Quantity of B/S in stock :   ....

13. Information on acceptability of variety by farmers/ consumers/industry : ....

14. Specific recommendation, if any for seed production :

....

15. Any other pertinent information : ....

16. Acknowledgement particulars about the submission of germplasm sample with NBPGR, New Delhi-12.

Signature of Head of the Institution

APPENDIX – III

Proforma for submission of the proposal for notification of crop variety under section 5 of the Seeds Act, 1966.

1. State :   ....

2. Crop :   ....

3. Name of the variety under which released or known :
   ....

4. Year of release :   ....

5.(a) Parentage with details of its pedigree :
   ....

   (b) Source of material incase of ...... & IE/EC No. Designation of introduction :   ....

   (c) Breeding method :   ....

   (d) Breeding objectives :   ....

6. State the varieties which most closely resemble this proposed variety in general characteristics :
   ....

7. Breeder/Institute responsible for maintaining breeders stock :   ....

8. Description of variety :   ....

9. Description of the parents of the hybrids. Is there any problem of synchronization, if yes methods to overcome it :   ....

10. Describe at least two identifiable and distinguishable morphological characteristics of the variety. In case of hybrid please describe also atleast two identifiable morphological characteristics of both the parents :
    ....

11. Maturity group (early, medium and late, where such classification exist) : ....

12. Disease and pest resistance (give details of any resistance to pests or disease including races) :
....

13. Recommended ecology : ....

14. Yield (in kg/ha) :

(*a*) Commercial product : ....

(*b*) Seed : ....

15. Current approximate percentage of area of the crop (kind) under this variety in the state : ....

16. Recommendation of the All India Workshop about the variety : ....

17. Acknowledgement particulars about the submission of germplasm samples with NBPGR : ....

Signature of the Chairman/Co-convenor

State Seed Sub Committee/Sponsoring Authority Name and Designation

2

PRINCIPLES OF SEED PRODUCTION

For development of a variety plant breeder tries to include as many genes as possible for different characters in his breeding material. When a variety is developed, it has several specific characters such as yield potential, straw stiffness, resistance to certain diseases and pests, earliness and nutritional qualities etc. When a farmer buys the seed of that variety he expects to buy not only a certain amount of seed but also all these specific characters which go along with that variety. A successful seed programme is able to supply a sufficient quantity of high quality seed at the required time, at reasonable cost and at the required place.

However, there is a long channel from the plant breeder's nursery until the seed reaches the farmers field. The plant breeder has very small quantity of breeder seed. The small quantity has to be multiplied over a number of generations to obtain sufficient quantity for commercial use. During this multiplication many things can happen which could change the characters of the variety so that the material which reaches the farmers is not the same as the variety bred by the breeder. More the number of seed multiplication, the greater the risk for deterioration of varietal purity. Now if there is considerable deterioration of varietal purity, the full value of the new variety will never be obtained.

Causes of Seed Deterioration

The deterioration of seed can be caused by different factors are as follows :

1. **Mechanical mixtures :** This is the most important and common source of variety deterioration during seed multiplication. This can occur with foreign material such as volunteer plants growing in seed field and sowing machines, threshers, cleaning and grading machines not properly cleaned before use and careless handling of seed during packaging, marketing and storage. Two varieties growing along side each other in the field are often mixed during harvesting and threshing operations.

2. **Natural out crossing :** Natural out crossing can be important source of varietal deterioration in sexually propagated crops. The extent of contamination depends upon the magnitude of natural-cross pollination. Crossing may takes place with the neighbourhood variety, with contaminants present in the seed field or with diseased plants, with undesirable types or off-type. Some times the segregates are not noticed in F_1 but could be seen in F_2. If aneuploids are existing they may cross with F_2 segregates and this will cause tremendous variability in subsequent generations. In self fertilized crops, natural crossing is not a serious source of contamination unless variety is made sterile or is grown in close proximity with other varieties. The natural crossing however, can be a major source of deterioration in cross-fertilized crops. The main factors, deciding the extent of contamination due to out crossing are the breeding system of species, isolation distance, varietal mass and pollinating agent.

3. **Residual variability :** It is quite difficult to secure a perfectly homozygous cultivar even in self-fertilized crops when it is developed. Residual variability may still be observed in the selected types. The cultivars continue to segregate for quite some times even after they have been released for cultivation. Under large scale cultivation the complex heterozygous throw out inferior and varying types that cause segregations. Especially in the case of disease resistance where the pathogens may show a large number of physiological races, such

deteriorations are frequent due to sudden multiplication pathogenic forms which under normal circumstances are not widely spread. Therefore, in case of breeding for disease resistance it is essential to test the strain for homozygosity for resistance under controlled optimum conditions of infection.

4. **Mutation :** Mutation with large effects can be identified and rogued. The rate of spontaneous mutation is very low in the population but this can not be ruled out during seed production. These micro mutants constitute real danger for genetic purity of the cultivar.

5. **Cytological instability :** When the meiotic division is not normal during cell division, gamete do not carry the normal chromosome number and thus they form aneuploid. These aneuploids throw tremendous variability in the population. Under these conditions some of the plants may become sterile which are more vulnerable to out-crossing and will ultimately produce more segregates in seed plots in subsequent generations.

6. **Developmental variations :** When the seed crops are grown in different agro-ecological conditions for several generations, developmental variations may arise because of differential growth response. The proportions of genotypes suited to such conditions is increased in the populations.

Methods of Maintaining Varietal Purity

It is extremely difficult to restore a variety from seed stock that has deteriorated. To make sure that the seed which reaches the farmers is true to the type, it is necessary (*i*) to continuously produce new breeder seed (*ii*) start new cycle of multiplication over and over again as long as the variety remains in the market.

Agrawal (1980) suggested the following steps to maintain genetic purity of varieties during seed production :

1. Use of only approved seed for seed multiplication.

2. Inspection and approval of seed plots prior to planting.

3. Field inspection and approval of crops at all critical growth phases.

4. Sampling and sealing of cleaned lots.

5. Growing of samples of approved stock for comparison with authentic stock.

Hartman and Kester (1968) recommended following steps for maintaining genetic purity of cultivars :

1. Provision of adequate isolation to prevent contamination.

2. Roguing of seed plot.

3. Periodic testing of varieties for genetic purity.

4. Growing crops only in areas of their adaptation to avoid genetic shift.

5. Compulsory certification of seed crop.

6. Adoption of generation system.

Sometimes it becomes necessary for the multiplication of varieties out side their areas of adaptation to maintain steady supply of required seed. This compulsion will be caused due to the weather limitation beyond ones control. The danger of genetic shift will be much greater in cases where the basic seed is multiplied outside their adopted area for more than three generations. To overcome this difficulty, it is advisable to multiply the basic seed in region of adaptation and subsequent production is taken for one or two generation out side the area of adaptation. Therefore, it is imperative to select the favourable area not only for high yield but also for free of various seed borne diseases which are not controlled by the seed treatment.

AGRONOMIC PRINCIPLES

Agronomic principles aid in production of high quality with maximum seed yield. The main agronomic principles are:

1. **Selection of Suitable agro-climatic regions :** A crop variety grown must be adopted to that agro-climatic regions. Regions of moderate rainfall and humidity are best suited for seed production than the regions of high rainfall and humidity. Poor pollination and pollen desiccation takes place due to high rainfall, humidity and high temperature leading to poor seed set and seedless fruits and poor quality seeds.

2. **Selection of Land :** The land selected should be fertile and able to fulfill the requirement of the crop. It should be levelled properly and with assured irrigation facility. It should be free from volunteer plants, weeds, other crop plants, soil borne diseases and insect pests. The same crop should not have been grown in the previous season. The plot should be feasible to isolate the plot from any contaminant as per the requirements of certification standards.

3. **Isolation distance :** It relates to spatial separation of seed crop from possible source of contamination during growing period. The isolation requirement of a seed crop depends on its mode of reproduction, pollen biology and mode of dispersal. Some crop species are reproduced vegetatively *e.g.,* potato, cassava, sugarcane etc. Second system of reproduction is by sexual process through seed. Sexual reproduction may be by self-fertilization or by cross-fertilization. A few crop species reproduced by apomixis which give seed like bodies and behave in all respect like normal seed. Self fertilization occurs when pollen from the anthers of flower is transferred to the stigma of the same flower and cross-fertilization occurs when pollen from one flower is transferred to the stigma of another flower to effect fertilization. In agricultural crop species pollen is transported either by wind or by insects. Those species, which rely upon wind usually, have erect flowering stem so that wind is able easily to disperse of the pollen and receptive stigma are not too encumbered by leaves. Species, which rely upon insect, have flowers, which are more elaborate, and which

attract insect with scents, coloured petals and good supply to nectar.

For the seed growers, the method by which cross pollination occurs is important. If wind is transporting agent, consideration needs to be given to the direction and velocity of prevailing wind in the area and account should be taken of the physical feature of land scape which might affect the wind. If insect have main agent for cross-pollination, it is essential to know which kind of insects are most efficient pollinator to be sure that seed producing area should be suitable habitat for that. In general cross-pollinated crops require greater isolation distance and self pollinated crops require less isolation distance.

4. **Selection of the variety and seed :** The variety selected should be very popular with the farmers and should have desirable quality like disease resistance, early flowering and grain quality. It should be really a high yielder adapted to the region. The seed should have been obtained from an authentic source of required class of seed, by observing validity period with certification seal and tag.

   Before the seed is planted, it should be treated with appropriate chemicals if not treated already. The seed treatment could be, with fungicides, bacterial in ocula-tion for legumes and for breaking dormancy due to hard seededness. Seed crop needs a lower seed rate when compare to commercial crop. Such spaced plant-ing facilitates inspection and roguing of the seed plot.

5. **Time and method of sowing :** Seed crop should be sown one week advance than the normal crop season. Some adjustment may be made to avoid pest and disease incidence. Sufficient moisture should be ensured during sowing to get better germination and optimum plant stand. Seed crops should invariably be sown in rows mostly by dibbling. Row planting helps in conducting

proper field inspections, facilitates roguing and plant protection measures.

For fibre and oilseed crops close planting in the rows is desired to branch more profusely at the top to produce more seed. In case of hybrid seed production, male and female lines should be planted across the wind direction in a recommended planting ratios. In case of highly cross-pollinated crops by insects, provision should be made for beehives in close proximity of the seed crop so that the seed setting will increase many fold.

6. **Control of weeds, diseases and insects :** It is a well known fact that the weeds compete with the main crop in respect of moisture and nutrition and reduce the seed yield in addition to contaminating the seed which is difficult to separate out later on. The weed may also serves as host for number of diseases. Preparation of a good seed field and following crop rotation are prerequisites for the elimination of the weeds to a greater extent. Proper weedicides can also be used for effective weed control.

   Infected crops with diseases and insects will give low yield and low quality seed. Systemic diseases, if not checked, produce diseased plants in the next generation such as loose smut of wheat. Some times the spores of the diseases are carried on the seed coat which can bring seedling diseases when planted with next season. Disease and insect infestation can be controlled by seed treatment to suppress seedling and seed borne diseases. Appropriate spraying schedule should be prepared and the infected plants should be removed from the seed field to check its spread to other plants.

7. **Roguing :** The act of removing undesirable individual plants from seed field is called roguing. Adequate and timely roguing is most important for quality seed production. The rogues which may differ from normal plant population may cause quick deterioration in seed stock by opportunities afforded for cross-pollination and

transmission of diseases etc. Off-types/rogues should, therefore, be removed at the earliest possible. Roguing may be carried at vegetative or pre-flowering, flowering stage or maturity stage of the seed crop. Pre-flowering stage roguing depends on the variation in plant height, colour of leaf, size, shape and auricle colour etc. Roguing at early stage may be necessary to remove virus affected plants.

In order to avoid genetic contamination, roguing at flowering stage is most important. Pollen shedders should be removed in hybrid seed production programme. Affected ear head should also be removed with great care and see that the spores do not spread.

At maturity roguing will avoid physical contamination with off-type and other contaminants.

8. **Fertility Management :** For the nutrition of seed crop, it is advisable to know and identify the nutritional requirements of seed crops and apply adequate fertilizer (nitrogen, phosphoros and potash) at proper time. Adequate, fertilizer application results in maximum yield, good seed quality and better expression of plant characters which ultimately facilitate roguing which helps in maintaining genetic purity as well.

9. **Irrigation :** Moderately dry regions are selected for seed production and thus it becomes essential to irrigate the seed crop during the growth stage. Number of irrigations and their interval depends on the crop duration, weather conditions and soil type. Most of the crops are irrigated at an interval of 10–15 days in light soils and 15–20 days in heavy soils. Care should be taken to avoid excess irrigation. Irrigation water should be provided as per the requirement of the crop and it should be stopped 2–3 weeks before seed maturity is reached. This will facilitate drying of plant and soil for easy harvesting. Excess water or prolonged drought adversely affect the germination and stand resulting in poor seed yield.

10. **Seed Certification :** It is a legally sanctioned system for quality control, seed multiplication and production and entails field inspection; pre and post control test and seed quality test to verify or check whether the seed crop meets the minimum field and seed standards. In seed certification field inspection is most important.

A. **Field Inspection :** The basic objective of field inspection is to ascertain that the seed being produced is of the notified variety not contaminated both physically and genetically beyond certain specified limits. The objectives are achieved by verifying that the seed crop is :

❑ Field meets the prescribed land requirement.

❑ Seed used for sowing of seed crop is from approved source.

❑ Provided with proper isolation or border rows in hybrid seed production.

❑ Planting ratio in hybrid seed production is followed.

❑ Properly rogued in confirmation with standards for different factors.

❑ True to varieties characteristics and no mechanical mixture, proper harvesting.

The field observations are then compared with a set of certification norms, specified for each crop in relation to different factors. Certification is only officially notified agency for the region concerned has the authority to perform the field inspection.

B. **Crop Stages for Inspection :** It is very difficult to verify the different factors affecting seed quality in the field in a single inspection as these factors do not occur at the same time and all may not affect at that growth stage. Thus the field inspection is required for all crops. The number of inspections and growth stage depends upon crop duration, mode of pollination and possibilities of contamination and nature of contaminating factors and stages of disease susceptibility.

In sexually propagated crops, the convenient stages of crop growth are classified as follows :

❑   Pre-flowering

❑   Flowering

❑   Post-flowering/at maturity

❑   Harvesting stage

In general, two, three and four inspections are made for self-pollinated crops, cross-pollinated and hybrid seed crops, respectively.

C. **Observations during field inspection :** Factors observed during field inspection vary among crops and growth stages. The source of genetical and physical contamination is a general factor and must be observed. Factors observed during field inspection are as :

❑   Off-types

❑   In separable other crop plants

❑   Objectionable weed

❑   Weeds

D. **Taking field counts :** In seed crops, it is not possible to examine all plants in the field. The number and method of count vary from crop to crop. For all crops, it is necessary to take a minimum of five counts upto 2 hectares of seed production area and an additional count for 2 hectares. The number of plants or heads that should make a count for different crops is as follows :

❑   For widely spaced and non-tillering crops like cotton and castor minimum number of plants in a count should be 100.

❑   Medium spaced and non-tillering crops like cowpea, black gram, green gram 500 plants per count should be considered. For others minimum 1000 plants/ear heads per count should be considered.

11. **Harvesting and Post-harvesting handling of Seed crop:** This is the most important operation which is needed to be taken up when the seed have reached physiological maturity which indicate cessation of metabolic activity from "Source" to "Sink". Maturation does not occur at the same moisture level. It differs with the crop, season and place. This is the stage when the seed recorded highest germination due to maximum accumulation of dry matter content in the seed. Early or delayed harvesting are not advisable. It is generally observed that the growers take up harvesting according to their convenience. Sometime it may be earlier than physiological maturity or delayed which may result in losses in yield and quality.

Harvesting time may be expressed based on : (*i*) Days after sowing (*ii*) Days after flowering or anthesis (*iii*) Days after 50% flowering (*iv*) Days after panicle or ear head emergence and (*v*) Days after 50% ear head emergence.

For most of the crops the moisture content of seed is the good indication of the optimum time to harvest. For soybean optimum moisture content at harvesting should be 12%, for wheat it should be 15–17%, for sorghum it is below 20% and maize crop can be harvested at 30–34 per cent before picked for drying.

Maximum care has to be taken during threshing to avoid mechanical mixing and injury to the seed in order to maintain the lustre and good appearance of the seed, threshing operation should be carried on a properly constructed and maintained threshing floor. If threshing is done by machine, it should be properly cleaned and adjusted in such a way that injuries to the seed are eliminated. When the harvesting and threshing performed by combine machine; it should be properly cleaned before use.

Drying and Processing : Normally the seed produced, after harvesting and threshing is subjected to sun drying to reduce the moisture content before carrying to seed processing plants. Direct exposure of the seeds to sunlight may affect its quality, therefore, it is necessary that seed may be dried under diffused sunlight in a shed. In some crops like paddy artificial drying is done by blowing dry air of a temperature varying from 70–85°F depending upon initial moisture of seed. It is always advisable to pre-clean the seed stock to remove dirt, chaff, soil particles etc. before drying operation is undertaken. Seed processing by using different model of grader should be done over prescribed sieve size or through gravity separator.

Seed treatment : Treatment of seed with prescribed doses of fungicides and insecticides should be done to control the carriage of disease causing pathogen and insects with the seed or their fresh entry into it.

Sampling and quality control : After cleaning and grading, the seed should be sampled with prescribed standard and tested in the seed testing laboratory to verify the different minimum seed certification standards.

Packaging, labelling and sealing : To ensure the delivery of seed produced through rigorous system of quality control, it is very important to use the right type of packaging material or container. If seed is to be packed in moisture–proof container, the moisture of seed should be brought down to a level of 6–8 per cent while for moisture pervious container, moisture content should not exceed 10–13%. For identification purpose each package should be labelled in prescribed manner and thoroughly sealed to prevent any tampering with the quality of seed.

Storage and Marketing : From the moment of maturity until planting, the seed is stored either on the plant or in a seed godown. By using proper storage condition, the rate of seed deterioration can be greatly slowed. Adequate precautions should be followed to prevent seed lot from deterioration while in transit or storage. Greater care should be taken for storage pests and diseases and the temperature and relative humidity of store. The moisture content of seed generally governs the length of time for which seed can be stored, coupled with the temperature in store.

Harrington and Douglas (1970) devised two simple rules for storage :

For every one per cent decrease in moisture content storage life of seed is doubled and for every 5°C decrease in storage temperature, the storage life is also doubled. The ideal temperature range for insect and fungal activity is 21°C to 27°C. Therefore, the storage temperature is much colder than 21°C as possible as required for long term seed storage.

The climate of seed storage location is important. Some locations provide satisfactory seed storage conditions naturally. One should remember that the seed godown better where the temperature plus relative humidity does not exceed 100. Based on this, Agrawal (1976) had studied few places in the country and identified them as a good moderate, poor, very poor places for short term seed storage. Similar attempts were also made by Bhattacharya and Chatterjee (1980) in West Bengal.

REFERENCES

Agrawal, P.K. (1976). Identification of suitable storage places in India on the basis of temperature and relative humidity conditions. *Seed Research*, 4 : 6–11.

Agrawal, R.L. (1980). Seed Technology. Oxford and IBH Publishing Company, 66, Janpath, New Delhi.

Bhattacharya, K.K. and Chatterjee, B.N. (1980). A note on the Identification of suitable seed storage sites in West Bengal. *Seed Res.*, 8 : 85.

Harrington, J.F. and Douglas, J.E. (1970). Seed storage and Packaging–Applications for India. NSC. Ltd. Rockfeller Foundation, India.

Hartman, H.T. and Kester, D.E. (1968). Plant propagation : Principles and Practices. Prentice Hall Inc., New Delhi.

3

CONCEPT OF GENETIC PURITY IN SEED PRODUCTION

Before 1960, the varietal improvement programme was state individual organiser scientist based. The integrated multidisciplinary approach to crop improvement was initiated in 1957 by launching the first All India Co-ordinated Project in maize crop. The system of official release of varieties came into being in October 1964 with the formation of Central Varieties Release Committee (CVRC) and State Varieties Release Committee (SVRC) in various states. The functions of CVRC were taken over by the Central Seed Committee in 1969 and notification of varieties was started. Along with the release and notification of improved varieties, the necessity of monitoring the quality of seed was also realized through seed certification.

The role of quality seed as a key resource to increase agricultural production cannot be over emphasized. Seeds of high yielding varieties not only act as catalyst for the utilization of other inputs, but their yield potential sets the limit for their utilization.

Quality Seed : It has been realized that seed is a primary input in agricultural industry and other inputs like fertilizer, irrigation, insecticides and weedicides will be responsive only if quality seed has been used. Quality seed has four parameters viz., physical purity, germinability, seed health and genetic purity.

Physical purity is the pure seed fraction from a seed lot and it is represented in terms of percentage.

Germinability indicates the capacity of seeds to germinate and emerge in the field with normal and vigorous seedlings. These two parmeters of quality seed determine the establishment of plant population per unit area. The population establishment is a foremost requirement for achieving the maximum possible yield. The poor physical purity and germinability directly affect the population establishment and hence poor yield.

Seed health, the third parameter of quality is related to seed borne diseases. Seed health affect the yield in two ways. Firstly, some of the diseased seeds do not germinate and hence decrease the size of plant population in the field. Secondly, some diseased seeds do germinate, emerge and establish in the field like normal plants. But in their advanced stage these diseased plants weaken and fail partially or completely to produce economic yield. The population of diseased plants increases in subsequent generations.

Genetic purity, the fourth parameter of quality seeds certainly does not affect the population establishment and requires a special mention. Genetic purity, generally, means that plants in a population of a variety are genetically identical and population is homogeneous. Genetic purity is one of the requirement of quality seeds under the Seeds Act.

These quality seeds are not, however, something only to be admired. They have to be multiplied, packed, stored and distributed, and planted in a scientific manner to make increased agriculture production a reality. Throughout this line the genetic purity of seeds along with other quality attributes have to be fully safeguarded otherwise, what are known as the seeds of hope may turn into seeds of frustration. It is therefore, imperative that we have a system whereby the genetic purity of seed lots could be checked effectively, and rather quickly to ensure that seeds moving in the commerce are tested for genetic purity before they are sold to the farmers.

Practical Constraints in Breeding Homogeneous Varieties

As pointed out earlier, that the degree of homozygosity

and homogeneity depends on breeding methodology employed by plant breeder. In self-pollinated species a variety develop by individual plant selection (pure line selection) may be homozygous and homogeneous for most of the loci. However, the level of homogeneity of the final population derived from a cross between distinct genotypes depends on the number of generations takes by single plant selection. Most of the cultivars are probably released after bulking of individual plant in 5th/6th generations and are therefore not as homogeneous as a pure line would be variety developed by line selection method would be much more heterogeneous.

The question arises whether a breeder really needs very high degree of homogeneity in a variety for achieving his ultimate goal *i.e.,* high yield? The level of homogeneity required in a variety is determined by the breeder, seed law enforcement agencies, farmers, and consumers. The breeder must be confident that genetic make up of a variety will not change during multiple generations of seed production. Farmers concern with uniformity is to have better harvestability and marketability. They are, generally, unconcerned about heterogeneity for the trait which do not affect their interest. The requirement of seed law enforcement agencies for varietal uniformity is for identification and estimating genetic purity.

On the other hand, breeders and geneticists have observed that the high degree of homogeneity jeopardise the stability and adaptability of variety. Their suggestion is to develop a population containing a number of co-adapted genotypes which are heterogeneously homozygous. However, final population should be uniform to level which does not affect the interest of farmers and consumers.

In self and often-cross-pollinated crops an acceptable heterogeneity is logical to harvest the benefits of interaction between genetically diverse lines. Breeder, generally, advises to maintain the variety during seed production, by bulking 200 or more plants or plant progenies. Otherwise, resultant population will not be true representative of original variety.

Causes for Genetic Deterioration in Varieties

Genetic purity (trueness to type) of a variety can deteriorate due to several factors during production cycle. The important factors of apparent and real deterioration of varieties are as under :

1. Developmental variations

2. Mechanical mixtures

3. Mutations

4. Natural crossing

5. Minor genetic variations

6. Selective influence of diseases

7. Technique of the plant breeder.

Maintenance of genetic purity during seed production

The method as suggested by Hartamann and Kester (1968) may be considered to maintain high levels of genetic purity during seed production.

1. Use of approved seed only during seed multiplications.

2. Inspection and approval of fields prior to planting.

3. Provide adequate isolation to prevent contamination by natural crossing or mechanical mixtures.

4. Field inspections at critical stages of crop growth for verification of genetic purity, detection of off-types, weed plants, abnoxious weeds and plants affected with seed-borne diseases.

5. Sampling of lots.

6. Periodic testing of varieties for genetic purity.

7. Avoiding genetic shifts by growing crops in areas of their adaptation.

8. Adopting the generation system.

In generation system the seed production is restricted to three or four generations only.

(*i*)   B/S $\longrightarrow$ F/S $\longrightarrow$ C/S

(*ii*)   B/S $\longrightarrow$ F/S I $\longrightarrow$ F/S II $\longrightarrow$ C/S

Generation System

Restricting multiplication to a limited number of generations from B/S (Breeder seed) is an important principle for maintaining varietal purity. Genetic purity in commercial seed production is often maintained through a system of seed certification.

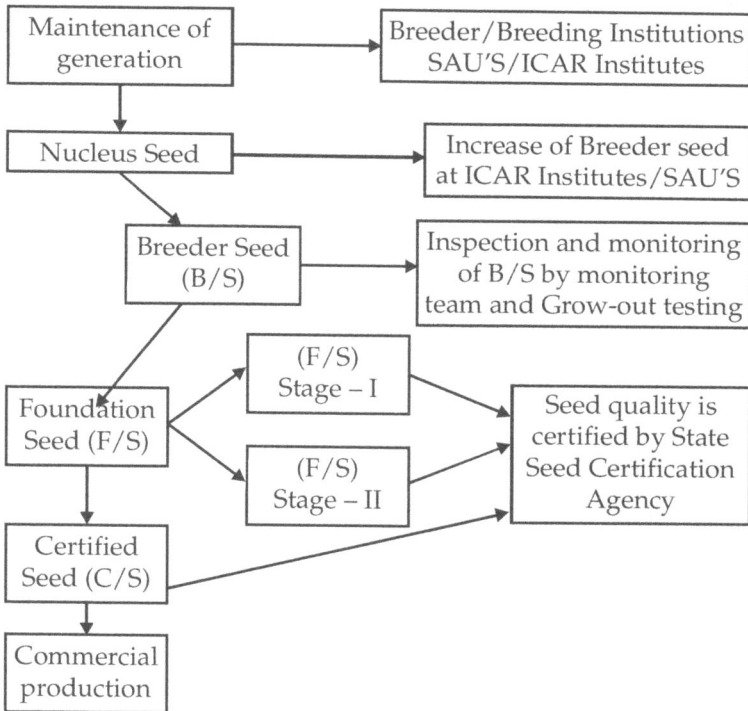

```
┌──────────────────┐         ┌──────────────────────────┐
│ Maintenance of   │────────→│ Breeder/Breeding Institutions │
│ generation       │         │ SAU'S/ICAR Institutes    │
└──────────────────┘         └──────────────────────────┘
         │
         ↓
┌──────────────────┐         ┌──────────────────────────┐
│ Nucleus Seed     │────────→│ Increase of Breeder seed │
│                  │         │ at ICAR Institutes/SAU'S │
└──────────────────┘         └──────────────────────────┘
         │
         ↓
      ┌──────────────────┐    ┌──────────────────────────┐
      │ Breeder Seed     │───→│ Inspection and monitoring │
      │ (B/S)            │    │ of B/S by monitoring      │
      └──────────────────┘    │ team and Grow-out testing │
         │                    └──────────────────────────┘
         ↓              ┌──────────────┐
┌──────────────┐       │ (F/S)        │      ┌──────────────────────┐
│ Foundation   │──────→│ Stage – I    │─────→│ Seed quality is       │
│ Seed (F/S)   │       └──────────────┘      │ certified by State    │
└──────────────┘       ┌──────────────┐      │ Seed Certification    │
         │         ────│ (F/S)        │─────→│ Agency                │
         ↓             │ Stage – II   │      └──────────────────────┘
┌──────────────┐       └──────────────┘
│ Certified    │───────────────────────────────────────→
│ Seed (C/S)   │
└──────────────┘
         │
         ↓
┌──────────────┐
│ Commercial   │
│ production   │
└──────────────┘
```

In genetic purity of seed is thus assured if the State Seed Certification Agency has approved the both seed crop and seed lot during seed processing. (On the basis of results obtained from Seed Testing Laboratory).

Seed Standard of genetic purity (%)

Foundation Seed	99%
Certified Seed	
(*i*) Varieties, composites and synthetics	98%
(*ii*) Hybrids	95%

GROW OUT TEST (GOT)

The monitoring of genetic purity of nucleus and breeder seed lots is very important which can be test by GOT. For this, a sample of appropriate size is drawn from the graded seed lots and grown as per the guide lines in the field to observe the genetic purity. The grow out test is conducted in the area where crop can express the maximum characters without any variation due to environmental influence. The crop is monitored throughout the growing season and off-type plants if any one tagged and counted at maturity to assess the genetic purity. The lots which are not of desired genetic purity should be rejected.

Sampling : The samples for grow out test are to be drawn simultaneously with the samples for other quality tests following standard procedure. The size of the submitted samples will be:

1000g – for maize, cotton, groundnut, soybean.

500g – for sorghum, wheat, paddy and species of other genera with seeds of similar size.

250g – Beta and species of other genera with seeds of similar size.

100g – for bajra, jute and species of all other genera.

Procedure : While raising the desired population, standard and recommended agronomic/cultural practices (*e.g.*, field preparation, size of the plot, row length, spacing between row, distance between plants, irrigation, fertilizers etc.) in respect of individual crops are to be followed both for the unknown sample and its control. Crop should be so grown that the genetical differences express themselves as clearly as possible.

In self-pollinated crops the individual of a cultivar may be theoretically identical whereas the individual of a cultivar in cross-pollinated species may not be genetically similar, but comprise a number of types. Therefore, it is easier to determine the cultivar purity in self fertilizing species than in cross-fertilizing species where the examination for greater part is based on the mutual comparison between the samples to be tested and the standard sample. Hence, it is essential to sow the various samples of the same cultivar in succession and standard samples are sown at suitable intervals (*e.g.,* one standard sample for every 10 samples to be tested). The size of plot, row length etc. will differ crop to crop.

Specifications for different crops :

S.N.	Crop	Row Length (m)	Distance (cm)			No. of replica-tions
			Between plants	Between rows	Between plots	
1.	Wheat, barley, Oat	6	2	25	50	2
2.	Pea, Cowpea	6	10	45	90	2
3.	Chickpea, Moong, Urd	6	10	30	60	2
4.	Maize	10	25	60	90	2
5.	Hybrid Cotton	5	10	45	45	2
6.	Paddy (early-medium)	6	15	20	45	2
	(Late and very late)	6	25	30	60	2
7.	Pearl millet	6	10	60	90	2
8.	Sorghum	6	10	45	60	2

The seed rate may be adjusted depending on the germination percentage of individual sample and the sowing may be done by dibbling. Subsequent thinning is not recommended. The test crop may be raised along with the control either in the areas recommended for the variety or in off-season nurseries. The authentic control sample from the originating plant breeder/breeding institute is to be maintained by the testing station/Agency following standard procedures. A minimum of two hundred plants from control sample will be raised along with the test crop.

Observations

All plants are to be studied keeping in view the distinguishing characters described for the cultivar both in the test crop as well as the control. Necessary corrections may be incorporated if the control *i.e.,* found to be heterogeneous. Observations are made during the full growing period, or for a period specified by originating breeding Institute and deviations from the standard sample of the same variety are recorded. At suitable development stage the plots are examined carefully, and plants which are obviously of other cultivar are counted and recorded.

Maximum Permissible Off-types (%)	Minimum genetic purity (%)	No. of plants required per sample observation
0.10	99.9	4000
0.20	99.8	2000
0.30	99.7	1350
0.50	99.5	800
1.00 and above	99.0 and below	400

The basis of the number of plants required for taking observations is dependent on maximum permissible off-types which are as follows :

Calculation, interpretation and reporting : Percentage of other cultivars, other species or aberrant found may be calculated upto first place of decimal. While interpreting the result, use of tolerance may be applied by using the reject table given below :

Reject number for prescribed standards and sample size

	Reject numbers for sample	Sizes of standard
	80	400
99.5 (1 in 200)	8	*
99.0 (1 in 100)	16	8
95.0 (5 in 100)	48	24
90.0 (10 in 100)	88	44
85.0 (15 in 100)	128	64

* *Indicates that the sample size is too small for a valid test.*

4

SEED PRODUCTION OF CEREALS

WHEAT (*Triticum aestivum* L. em. Thell)

Wheat is the World's second most important staple food crop of family Gramineae. Its cultivation is favoured by long cool, moist weather followed by dry and warm weather. Among different wheat species, *Triticum aestivum* (hexaploid), bread wheat, has the major share of its cultivation. *Triticum durum* (tetraploid), durum wheat, also has considerable area under cultivation. Two other species *Triticum monococum* and *Triticum timopheevii* are also cultivated in some areas of the world but they have a little economic importance.

Important Diagnostic Characters

To facilitate roguing of the off-type plants from a seed field, the knowledge of diagnostic characters of a variety is very important. These diagnostic characters should have high heritability so that does not change in different environments. The important diagnostic characters are presented in Table 1.

Important Steps Involved in Seed Production

Land Requirement : The land that is to be used for seed production should be free from volunteer plants. In the previous year the wheat crop should not be taken in the field where seed production is to be taken.

Isolation Distance : Wheat is a self-pollinated crop and the seed production plot should be isolated from other wheat field by a minimum of 3 meters distance. In case of neighbouring fields are infected with loose smut, the seed production plot should be isolated from them by a distance of 150 meters, if the infection is in excess of 0.1 per cent in case of foundation seed production and 0.5 per cent in case of certified seed production.

Table 1 : Diagnostic characters of wheat

Vegetative phase	T. aestivum	T. durum
Early leaf width	Broad	Narrow
Late leaf width	Narrow	Broad
Leaf surface	Smooth	Rough
Leaf colour	Medium dark	Light to dark
Ligules	Rudimentary	Rudimentary
Auricles	Developed	Fin Shape
Hairs on auricle	Hairy	Non hairy
Reproductive Phase		
Ear	Very lax to very dense	Mid dense-dense
Spikelet	Three spikelets	Three spikelets
Awn	Short to medium	Long
Grains	No hump	Prominent hump

Seed Source : For foundation seed production Breeder seed is to be required. B/S should be purchased from ICAR institutes/ SAU's. For certified seed production F/S is required. At the time of purchase of seed the bag should be properly stitched, tagged and sealed. The validity period of the seed lot is to be high. The seed should be treated with vitavax @ 2.5g/kg of seed to prevent loose smut. The seed rate per hectare ranges from 80–90 kg and row to row spacing is 23 cm.

Fertilizer Application : It is always recommended to apply the fertilizer based on soil test results. The fertilizer supplying 120 kg. Nitrogen, 60 kg P_2O_5 and 40 kg. K_2O per hectare. Half of the nitrogen and entire quantity of phosphorus and potash is applied at the time of sowing and remaining half nitrogen is applied when the crop is 30–35 days old. The fertilizer should not come in contact with the seed, therefore, it is placed about 5 cm away from the seed row.

Irrigation : Irrigate the crop at crown root initiation proba-bly at 25–30 days after sowing. Other irrigations should be given at tillering, flowering, boot leaf stage and grain filling stage.

Roguing : Timely removal of off-types is essential to maintain the purity of the seed. These can be identified by the differences in colour, susceptibility to disease, plant height, variation in the ear head, smutted plant and early flowering plant. Extreme care should be taken of smutted plants. While removing the plants the ear head should be covered with the paper bag and uprooted without allowing the spores to fall. These plants are collected in a gunny bag and burnt or buried. After completion of the flowering, roguing may be taken up again to spot such plants which have escaped your attention during first roguing. Final roguing can be taken up when it is possible to remove the Off-type on the basis of differences in colour of the ear head, awns and ear head type. At this stage or prior to harvest the objectionable weed plant (Hiran Khuri *Convolvulus arvensis*) may be removed.

Harvesting and Threshing : Extra care is needed to avoid mechanical mixing that is likely to occur during harvesting and threshing. If harvesting and threshing is done by combine, it should be properly cleaned. Before harvesting of seed field it is advisable that 1 meter strip around the seed field should be harvested and keep it separately. Again clean the combine and then start the harvesting of the seed plot.

Bagging and Storage : After harvesting and threshing seed should be properly cleaned and dried. The moisture content of the seed should not be more than 12 per cent.

Field Standards	F/S	C/S
Isolation distance (mini.)	3m	3m
Other varieties (max.)	0.05	0.20
Objectionable weed plants (max.)	–	–
Diseased plants (loose smut) max.	0.10	0.50

Seed Standards	F/S	C/S
Physical purity (mini.)	98%	98%
Inert matter (max.)	2%	2%
Other crop seeds (max.)	10/kg	20/kg
O.D.V.	–	–
Weed Seeds (max.)	10/kg	20/kg.
Objectionable Weed Seeds (max.)	2/kg	5/kg
Karnal Bunt infection (max.)	0.10	0.50
Germination (minimum)	85%	85%
Moisture (max.)	12%	12%

Barley (*Hordeum vulgare* L.)

Barley originated from middle East and is widely grown in countries of diverse climatic conditions. It is used for grain, feed and fodder. It is tolerant to alkali, drought and frost.

Land Requirement : Land which is to be used for seed production should be free from volunteer plants. In the previous year where barley crop was grown should be avoided for seed production in that field.

Methods of Sowing : Seed rate of 75 kg/hectare is recommended for sowing. Sowing should be done in rows at 23 cm apart. If sowing is to be done by seed drill it is insured that seed drill should be properly cleaned before use.

Fertilizer Requirement : 60:30:30 Nitrogen, phosphorus and potash (kg/hect.) is optimum dose of fertilizer should be applied in seed crop. Half dose of nitrogen and entire dose of phosphorus and potash should be applied as basal dressing. Remaining half quantity of nitrogen should be given after first irrigation (30–35 days after sowing).

Irrigations : First irrigation after 30–35 days of sowing and 2nd irrigation should be given at milking stage.

Seed Treatment : If seed is not treated then treat the seed with Thiram @ 2.5 g/kg. seed before sowing in the field.

Isolation Distance : Barley is a self-pollinated crop and requires 3 meter isolation distance for foundation and certified seed production. In case of neighbouring fields of barley are infected with loose smut, the seed production plot should be isolated from them by a distance of 150 meters, if the infection is in excess of 0.1 per cent in case of F/S and 0.5 per cent in case of C/S production.

Seed Source : The seed should be obtained from the approved source of appropriate stage. If we are going to produce F/S of any variety (notified) of barley then we need B/S of this variety and it should be purchased from ICAR institutes or State Agricultural Universities.

Roguing : This operation must be carried effectively for quality seed production. All the off-type plants and volunteer plants should be removed from the seed plot before flowering. They must be pulled out to prevent growth. Loose smut and covered smut infected plants should be rogued out very carefully so that the smut spores can not be fall in the seed field.

Harvesting and Threshing : After final inspection of seed plot by the Certification Agency, seed crop should be harvested and threshed in such a manner so that there may not be any type of contamination in seed.

Field Standards	F/S	C/S
Isolation distance (mini.)	3m	3m
Other varieties (max.)	0.05	0.20
Diseased plants (Loose smut max.)	0.10	0.50

Seed Standards	F/S	C/S
Physical purity (mini.)	98%	98%
Inert matter (max.)	2%	2%
Other crop seeds (max.)	10/kg	20/kg
Weed seeds (max.)	10/kg	20/kg
Objectionable Weed seeds (max.)	2/kg	5/kg
Germination (mini.)	85%	85%
Moisture (max.)	12%	12%

Paddy (*Oryza sativa* L.)

Land Requirement : The soil should be clay loam to clay with a pH value of 6.5 and land should be well levelled.

Isolation : Being a highly self pollinated crop, an isolation distance of 3 meter from other paddy field is sufficient for quality seed production.

Preparation of Land for Transplanting : The land is prepared by ploughing to get fine tilth. The ploughing in wet condition in standing water 3 or 4 times will stir up and puddle the soil thoroughly. Organic or green manure, if used are ploughed in 15 days before transplanting. To hold irrigation water evenly, the levelling of the field is essential.

Fertilizer Application : The total requirement of the crop is 100–120 kg nitrogen, 50–60 kg of phosphorus and 50–60 kg potash per hectare. Apply the entire phosphorus and potash and half of the nitrogen just before puddling of the field. Out of remaining dose of nitrogen, 1/4 should be given at tillering stage and remaining 1/4 top dressed at panicle initiation stage.

Transplanting : Fifteen-twenty days old seedlings should be transplanted in rows with a spacing of 20 cm between rows with 10 cm spacing between the plants. At least 2–3 seedlings per hill should be transplanted.

Water Management and Inter Culture Operations

From the day of transplanting, the water level in the field should be maintained at 2.5 cm height for first 10 days. This water level can be increased to 5.0 cm till maturity. From tillering stage to panicle initiation water level should not be changed.

Hand weeding may be taken up two or three times before the crop starts heading. For the control of broad leave weeds 2, 4-D can be sprayed after 20–25 days after transplanting @ 2.5 kg dissolved in 750 lit. of water/hectare or 2, 4-D Ethyl Aster 5% granule @ 15 kg/hect. when the crop is one week old or Butachlor 5% @ 30 kg/hect.

Plant Protection Measures

For the control of Brown plant hopper spray the crop with 0.5 per cent Sevin solution/hect. or dust with 5 per cent Sevin @ 25 kg/hectare. There should be 3 cm water level in the field at the time of treatment.

Stem borer is controlled by spraying Endrin 25 EC@ 1.25 lit./hectare.

For control of Gundhi Bug, dust 25 kg/hectare 5 per cent BHC dust.

Blast can be controlled by spraying Benlate 150g/hect. in 250 lit. of water.

Bacterial leaf blight can be controlled by spraying 75g Agrimycin 100 + 500 g copper oxychloride in 500 lit. of water/ hect. This should be repeated 3-4 times at an interval of 10–12 days.

To control against seedling blight and Foot Rot, seed should be treated with Benlate @ 2 g/kg. seed. Khaira disease can be controlled by spraying with $ZnSO_4$ @ 5 kg/ $ZnSO_4$ + 2.5 kg. Calcium hydroxide in 1000 lit. of water/hect.

Roguing

Off-types should be removed from the seed field. Roguing must be done before flowering, during flowering and at maturity stage of the seed crop.

Wild rice plants, stem borer infested plants and diseased plants affected by Tungru virus and false smut should be removed from the seed field.

Harvesting, Threshing and Drying

When the crop is matured and finally approved by the seed certification personals harvesting can be done manually or with combine. Just after harvesting and threshing the moisture content should be brought down 13–14 per cent. During harvesting and threshing care should be taken so that seed can not be contaminated with other varieties. The seed should be made free from chaff, dust, empty husks, light grains and soil particles.

Field Standards	F/S	C/S
Isolation (mini.)	3m	3m
Inseparable other plant (max.)	0.01	0.05
Plants affected by false smut	0.10	0.50

Seed Standards	F/S	C/S
Pure seed (mini.)	98%	98%
Inert matter (max.)	2%	2%
Other crop seed (max.)	10/kg	20/kg
Total weed seeds (max.)	10/kg	20/kg
Objectionable weed seed (max.)	2/kg	5/kg
Germination (mini.)	80%	80%
Moisture	13%	13%
Rice bunt (max.)	0.1%	0.5%

HYBRID RICE

To obtain the benefits of hybrid rice cultivation it is essential to produce and supply required quantity of quality seeds at reasonable price to the farmers.

Season and Location for Seed Production

In India Rabi season (November) has been found to be better than other season (June-October) for hybrid rice seed production. Seed yield is obtained almost double in rabi season compared to kharif season. In Northern India rice is not grown during rabi season, due to very low temperature, so the large scale seed production is to be taken up in South India during rabi season. The field should be selected based on the following seasonal condition during flowering for higher seed set :

❑ Daily mean temperature should be 25–30°C.

❑ Relative humidity should be 70–80 per cent.

❑ Difference in day and night temperature is about 8–10°C.

❑ Sufficient sun shine and moderate wind velocity.

❑ No continuous rainfall for 10–12 days during peak flowering period.

Selection of Field

The field selected for hybrid rice seed production should be fertile with good irrigation and drainage facilities. For achieving synchronous flowering, a homogeneous plot with even topography is very much desired. The field should not be infested with serious pests and diseases. The field should be free from volunteer plants.

Isolation

Isolation of the seed production plot is necessary as rice pollen which is small in size can travel longer distance through winds. Isolation can be provided by different means :

1. **Space Isolation :** A space isolation of 100 meter would be satisfactory for hybrid seed production which implies that with in this range no other rice varieties should be grown except the pollen parent. For 'A'-line multiplication it is safer to have an isolation distance of about 500 meter.

2. **Time Isolation :** When it is difficult to have space isolation; a time isolation of over 21 days would also be effective. It means that the heading stage of the parental lines in hybrid seed production plot should be 21 days earlier or late than that of other varieties grown within the vicinity.

3. **Barrier Isolation :** In some places, the natural topographic features such as mountains, rivers, forests, can serve as the most effective barrier. A crop barrier with maize, sugarcane, sesbania (Dhaincha) covering a distance of 3 m would also serve the purpose of isolation. Artificial barrier with polythene sheets of about 2m height can

also be used for small scale seed production. However, the most ideal locations are the areas covered with hills and mountains which act as natural barriers.

Row Ratio, Row Direction and Planting Pattern

In hybrid rice seed production the seed parent and pollen parent are planted in a certain row ratio at a certain spacing. The row ratio and spacing of pollen parent and seed parent have a distinct effect on the hybrid seed yields.

The row ratio or row proportion refers to the number of rows of the male parent (R-line) to female (A-line) in a seed production plot. In hybrid rice seed production plot the recommended male (R) to female (A) row ratio is 2:8. However, the ratio may vary from region to region, depending on weather, management and parental lines. R and A lines can be planted in several row ratios of 2:8, 2:12, 3:10 etc.

Factors Influencing Row Ratio : The ratio of pollen parent (R-line) to seed parent (A-line) is determined by the characteristics of the parental lines.

❑ Plant height of the pollinator

❑ Growth and vigour of the pollinator.

❑ Size of the panicle and amount of residual pollen.

❑ Duration and angle of floret opening in CMS line.

❑ Stigma exertion of CMS line.

To facilitate out crossing, it is better to adjust the row direction nearly perpendicular to the wind direction prevailing at the time of flowering.

Differential Seeding

The parental lines differing in their growth duration can be sown on different dates in the nursery beds so that they come to flowering at the same time in the main field. This is called the differential seeding or staggered sowing. The difference in seeding dates of pollen and seed parent is called as seeding interval. The seeding interval between parental lines may vary from 2–3 days to 45 days. It is rare that both parents of a given combination take the same number of days to flowering.

Special Techniques

(*a*) **Flag Leaf Clipping :** Enhancing out crossing rate is one of the key factors to increase the seed yield. Normally the flag leaves are relactant longer than the panicles and they come in the way of easy pollen dispersal thus affecting the out crossing rate. Flag leaves should be clipped when the main culms are in booting stage. Flag leaf clipping helps in uniform pollen movement and wide dispersal of the pollen grains to get higher seed set. Flag leaf is very important and hence utmost care is needed while clipping the flag leaves. At booting stage, the upper leaves of the plants are held properly and they are cut with the help of a sharp sickle in such a way that 1/2 or 2/3 of the flag leaf is removed. The cut is to be given at a level just above the flag leaf joint.

(*b*) **Gibberellic acid (GA$_3$) :** Most of the WA based cms lines have imperfect panicle exsertion, with the result 10–15% spikelets are enclosed in the flag leaf and not available for out crossing. This is one of the drawbacks of these cms lines. In view of their good stability and usability, it is necessary to depend on these lines and find ways and means of improving of their panicle exsertion. Gibberellic acid is an efficient and effective growth hormone which stimulate the cell elongation. This chemical can be used to enhance panicle exsertion of cms lines based on WA system. GA$_3$ also increases the duration of floret opening, increases stigma receptivity and useful in widening the flag leaf angle.

Timing : In hybrid rice seed production plots, 5–10% panicle emergence (based on tillers) stage is most appropriate for the first spraying (40%). The remaining 60% of GA_3 has to be sprayed the next day itself. Morning (8–10 am) and evening hours (4–6 pm) are ideal for spraying. Spraying should be avoided on cloudy or rainy days and high wind velocity period.

Dosage of GA_3 : Several studies indicated that 45–60 g is optimum. However, this dose can be still lowered by the use of ultra low volume Sprayers. GA_3 can be partially substituted by other chemicals like Urea (2%), Boric acid (1.5%) and NAA (50 g/hectare) in order to reduce the cost of seed production.

(c) **Supplementary Pollination** : Rice is basically a self pollinated crop and hence there is a need to go for supplementary pollination in order to enhance the extent of out crossing. It is a technique of shaking the pollen parent so that the pollen is shed and effectively dispersed over the 'A'-line plants. It can be done either by rope pulling or by shaking the pollen parent with the help of two bamboo sticks. Timing and frequency of supplementary pollination is very important. The first supplementary pollination should be done at peak anthesis time *i.e.,* when 30–40 per cent of the spikelets are opened. This process is repeated 3–4 times during the day at an interval of 30 minutes. It has to be done for 7–10 days during the flowering period.

Agronomic Manipulation

Establishment of a good, vigorous and healthy crop is a pre-requisite to obtain high seed yield. Fertilizer dose of 120 : 60 : 60 N, P_2O_5, K_2O kg/ha. is suggested. Then nitrogen is given in 3–4 split doses and potassium in two split. More split of nitrogen to male parent will prolong pollen supply.

There should be good irrigation and drainage management of seed production plots. Excess standing water promotes early flowering while draining the water delays flowering. Male parent responds more than female parents to water regulation.

Prediction of Flowering and Adjustment for Synchronization

Inspite of accurately determining and adopting differential seeding, there will be minor differences in flowering of parental lines due to seasonal and weather fluctuations and differences in crop management practices. Hence prediction of flowering needs to be done during growth of seed crop and appropriate measuring for adjustment need to be adopted.

Flowering can be predicted based on the panicle initiation and its subsequent stages of growth. Panicle initiation takes place approximately 30 days before flowering about 30–40 days after transplanting. There are 10 stages of panicle primordium development and one should have good knowledge of these stages, and these stages can be observed by splitting open the main stem of a plant. The first three stages are the most important in predicting and adjustment of flowering are very small and tiny (0.2 – 1.2 mm) and need to be observed with a magnifying lens.

Adjusting of flowering for 2–3 stages or 4–8 days can be done by adopting appropriate measures during first three stages of panicle development. For delaying flowering quick releasing nitrogen fertilizer (2% urea) can be sprayed on earlier parent. For hastening flowering, 1% solution of phosphatic or potassium fertilizer is sprayed on late parent. At 25–30°C, stigma is receptive for 4–5 days while pollen is viable only for 4-5 minutes. Hence when pollen grains are released, maximum number of stigma should be in receptive stage. At higher temperatures, receptibity of stigma and viability of pollen grains reduce drastically.

Roguing

To remove the off-type and volunteer plants from the CMS and restorer population in the seed production plots. Roguing can be done at any time during the crop stage. Off-type rogues can be removed whenever they are identified–earlier the better. The most important stages for roguing are at maximum tillering, at flowering and just before harvesting.

Harvesting, Threshing and Drying

In all these operations it is most important that there are no chances of mixing of seed from pollen parent and seed parent at any stage. Male parent is harvested first and taken away for threshing. Then after a critical examination of female rows for off-type and their removal, seed parent is harvested and threshed separately.

Field Standards	F/S	C/S
Fields of other varieties including commercial hybrid of the same variety	200 m	100 m
Fields of same hybrid not conforming to varietal purity requirements for certification	200 m	100 m
Off-types in seed parent	0.05	0.20
Off-types in pollinator	0.05	0.20
Pollen shedding ear heads in seed parent	0.05	0.10
Objectionable weed plants	0.01	0.02

Seed Standards	F/S	C/S
Pure Seed (mini.)	98%	98%
Inert matter (max.)	2%	2%
Huskless seeds (max.)	2%	2%
Other crop seeds (max.)	10/kg	20/kg
Other distinguishable varieties (max.)	10/kg	20/kg
Total weed seeds (max.)	10/kg	20/kg
Objectionable weed seeds (max.)	2/kg	5/kg
Seed infected by Rice bunt (max.)	0.10%	0.50%
Germination (mini.)	80%	80%
Moisture (max.)	13%	13%

MAIZE (Zea mays)

Maize is one of the major agricultural crop being only next to wheat and rice in importance. It is highly cross-pollinated being monoecious in nature which makes it easy for production of hybrid seeds. It is used as food, feed and for industrial purpose. It can be grown under rainfed and irrigated conditions.

Land Requirement

Loamy soil with high organic matter with neutral pH is the best for maize crop. The land should be free from volunteer plants.

Isolation

To prevent foreign pollen contamination, the seed field must be isolated at least by 400 meters for F/S and 200 meters for C/S production, respectively.

Spacing

Row to row spacing – 75 cm

Plant to plant – 20 cm

Fertilizer

120–150 kg Nitrogen, 60–75 kg phosphorus and 40–50 kg potash/hectare. In case of zinc deficient soils, incorporate 25 kg zinc sulphate/hectare. Apply 1/3 quantity of nitrogen and the entire P_2O_5 and K_2O before sowing as basal or at the time of sowing. Remaining 2/3 nitrogen top dressed 3–4 weeks after sowing.

Interculture

Interculture operations are necessary for good aeration and removal of weeds.

Roguing

All the off-type plants must be removed from the seed plot. Roguing of off-type plants should be completed before pollen shedding. Malformed and diseased plants should be rogued from the field.

Harvesting, Sorting and Drying

The crop should be allowed to dry to 15% moisture content before harvesting. Off-type maize ears showing different colour, texture and disease should be sorted out before drying. At this stage approval has to be sought for shelling from seed certification agency and in their presence shelling and grading is done.

Seed Standards

Pure seed (mini.)	98%	98%
Inert matter (max.)	2%	2%
Other crop seed (max.)	5/kg	10/kg
Weed seeds	–	–
Germination (mini.)	80%	80%
Moisture	12%	12%
Off-types (max.)	0.2	0.5
ODV	5/kg	10/kg

HYBRID MAIZE

Depending upon the number and the arrangement of the parental lines, the following types of hybrids are possible :

1. **Single cross :** Cross between two inbred lines, (A × B).

2. **Three way cross :** First generation seed of a cross between a single cross and an inbred line, (A × B) × C.

3. **Double cross :** First generation seed resulting from a cross between two single crosses, (A × B) × (C × D).

4. **Top cross :** First generation seed resulting from a cross between an inbred and an open pollinated variety, A × OPV.

5. **Double top cross :** First generation seed resulting from a cross between a single cross and an OPV, (A × B) × OPV.

6. **Multiple cross :** It is combination of more than four inbred lines.

7. **Composite varieties :** Advance generation of multiple crosses between selected varieties of diverse origin. Here the base population used in compositing will be heterogeneous.

8. **Synthetics :** Advance generation of multiple crosses between selected inbred lines.

The various steps in hybrid seed production are :

1. Maintenance of inbred lines (parental lines)
2. Production of single crosses
3. Production of double cross seed

Maintenance of Inbred Lines

Maintenance of inbred lines should be carried by adopting the following method. All the agronomic practices followed for the production of OPV should be followed except that the seed rate per hectare is reduced. Harvest and post harvest practices are similar to OPV including plant protection measures.

Isolation

(*i*)   400 meters from any maize with same kernel colour and texture.

(*ii*) 600 meters from maize of the different kernel colour and texture.

(*iii*) 400 meters from maize of the same variety not conforming to the purity requirement for certification.

Roguing

Every effort should be made to keep the inbred line genetically pure by adopting vigorous roguing of off-types before pollen shedding. Roguing can be started from the knee high stage of the crop by identifying the off-type through their morphological characters such as height, leaf size and shape and silk characters etc. Stalk rot affected plant should be rogued finally. Suckers and diseased plants should also be removed.

Single Cross Production

Land requirement and other cultural practices remain the same as followed for the seed production of inbred lines.

Isolation Distance

Same isolation as in case of inbred line seed production. Differential blooming dates are permitted for modifying isolation distance, provided 5 per cent or more of the plant in the field, with in the prescribed isolation distance, are shedding pollen.

Seed and Sowing

Seed of inbred lines should be procured from the source approved by the certification Agency.

The single cross seed is produced on inbred lines. In a crossing block two required inbred lines are sown. One line is used as female parent and the other is used as male parent. The planting ratio is 2:4 between male and female line.

Seed Rate

10 kg/hect. (female)

5 kg/hect. (male)

Datasselling

The seed parent should be dataselled before anthers shed the pollen so that the silk of the line would receive the pollen from the desired parent. From the time the first tassel appearance, datasselling work has to be continuously carried without fail until the last tassel on female line is removed.

After 50 days of planting, watch the crop in female rows before the appearance of tassel. Datasselling should be done everyday when the tassel is fully out of the leaf sheath but before anthesis. Datasselling should be done carefully so that the portion of the tassel may not remain on the plant which will contaminate the seed parent.

Commercial Hybrids

The production of hybrid seed is meant for its distribution to the farmers for commercial cultivation. The hybrid seed is produced on a high yielding single cross (A × B) which is used as female parent. The hybrid seed can be a three way cross, double cross or double top cross. Here the pollinator parent could be a single cross, OPV or inbred line. Female parent which is the product of single cross is subjected to datasselling before pollen shedding to ensure cross pollination from the desired male parent. The seed obtained from the female parent is hybrid seed.

Seed Rate

Female parent – 10 to 12 kg/hect.

Male parent – 4 to 5 kg/hect.

There will be 6 rows of female for every two rows of male inside the seed production plot. In addition, there will be two rows of male in the border around the field. The male and female parents are sown simultaneously.

Isolation

200 meters for maize with same kernel colour and texture.

300 meters for maize with different kernel colour and texture.

200 meter for maize of the same variety not conforming to the varietal purity.

Seed Standards	F/S	C/S
Pure seed (mini.)	98%	98%
Inert matter (max.)	2%	2%
Other crop seed (max.)	5/kg	10/kg
Weed seeds	—	—
Germination (mini.)	80%	90%
Moisture (max.)	12%	12%
Off-types (max.)	0.2%	0.5%
Max. shedding tassel	0.5%	1.0%
ODV	5/kg	10/kg

Hybrid Sorghum

For the production of certified seeds of hybrid sorghum, a crossing block with male sterile line (A-line) and pollinator line (R-line) has to be established.

Commercial seed production of hybrid sorghum must be carried out in well organised manner. Cytoplasmic genetic male sterility system is utilized in the commercial seed production. This involves the cycle of two years and three steps :

First Year : Simultaneous production of :

(*i*)   Foundation seed production of A-line (A × B).

(*ii*)   Foundation seed production of B and R-lines.

Second Year : Involves the additional step of certified seed production of hybrid (A × R).

Two isolation fields are needed in the first year for production of F/S of A-line and R-line separately in isolation. In subsequent years, three isolation distances are needed with addition of certified seed production of hybrid.

Isolation

Fields of other varieties of grain and dual purpose sorghum	200 meters.
Fields of the same variety not conforming to varietal purity requirement for certification	25 m
Johnson grass	400 m
Forage sorghum	200 or 400 m

400 meters from grossy sorghum.

(Hightillering and grossy panicle and 200 meters from the sorghum meant for grain and fodder, but used primarily for fodder.

Foundation Seed Production of Male Sterile Line (A-Line)

Male sterile line (A-line) is a female line, which does not produce fertile pollen, thus can not be produce seed, and multiply by it self. Hence, it has to be multiplied with the help of maintainer line (B-line), which is a isogenic line of male sterile line (A-line), but has a fertile pollen in its anther sacs. Pollen from this maintainer line fertilizes the male sterile line and at the same time maintains the sterility in male sterile line. Hence seed produced on male sterile line with the help of maintainer line will produce sterile plants which are needed to produce hybrid seeds in next year by pollination with the desired R-line.

Large scale production of male sterile lines is carried out by growing the A-lines and its corresponding maintainer B-line together in isolation plot. The isolation distance required for A × B production fields is the same as that recommended for A × R production *i.e.,* 300 meters. Production procedures are similar to A × R production except that the R-line is replaced by the B-line. A ratio of 4A : 2B or 6A : 2B rows is maintained and borders of the field are sown with B-line. Pollen produced by B-line fertilizes the male sterile plants (A) and the seeds produced thus gives rise to A-line, which will be sterile when grown again. As these are isogenic line except fertility, they flower at about same time.

Roguing in A-line seed production plots should be more stringent, because A and B plants can not be distinguished after flowering. The shedders in the A-line rows must be identified and uprooted each morning during the flowering period. Utmost caution must be exercised in labelling and harvesting A-line rows and B-line rows. The B-line rows are harvested first, followed by the A-line rows. Purity of the A-line is very important and any lapses can lead to huge losses of time and money spend in roguing the hybrid seed (A × R) production plots in the next generation. Since the A and B-lines are isogenic lines, grow with equal height and exhibit synchronous flowering, thus there is no problem of synchronization either for flowering or plant heights. Therefore, the seed yields on the A-line are relatively better in A × B seed production plots than in A × R (hybrid) production plot. Seed of the B-line harvested from the A/B production plots might be reused for the next generation, depending on the seed laws of the country.

Foundation Seed Production of B and R-Lines

Breeder seeds of B-line and R-lines are to be obtained from concerned breeder. Both maintainer (B-line) and restorer (R-line) have self-fertile bisexual florets and are pure lines. The seed multiplication plot of R-line is sown in an area isolated by a radius of >200 m distances from other sorghum cultivars. If Johnson grass or any other forage sorghums are growing in the vicinity, an isolation distance of 400 m is recommended for the multiplication of R-lines. Any plant in the R-line plot appearing different from the true R-types, for any character should be uprooted, or rogued before anthesis. Although the process of roguing or removal of off-types, starts soon after the seedling stage, the boot leaf and panicle emergence stages are most critical, because detection of off-types is easier during these stages. Off-types that escape detection during the flowering stage should still be removed before harvest to minimize contamination.

The maintainer or B-line, which is also a self-fertile, can be multiplied in an isolated plot in the same manner described above for the R-line.

Certified Seed Production of Hybrid (A × R)

During second year *i.e.,* after multiplication of required quantities of F/S of A and R-lines, the third step is certified seed production of the hybrid seeds. Hybrid seed is produced by growing the designated male sterile (A) line and restorer R-lines of the hybrid together in an isolated field and allowing cross pollination. An isolation distance of >300 m is generally recommended for hybrid seed production, although a distance of >400 m necessary if Johnson grass and other forage or grassy sorghum types are growing in the vicinity. Row spacing can vary from 45–90 cm. Male sterile (A) and restorer (R) lines are sown in alternate strips of rows, normally in the ratio of 4A : 2R or 6A : 2R depending on the local experience of success and the ability of the R-line to disperse pollen. The borders on all four sides of the hybrid seed plot are sown with restorer (R) line to ensure an adequate supply of pollens.

Roguing

Roguing should be carried out in the hybrid seed production field regularly as soon as the crop commences flowering. Apart from off-types, pollen shedders can be a problem in A-lines. Such plants can only be identified at anthesis and should be uprooted immediately. Shedders can also arise from partial break down of sterility in A-line under high temperature (>38°C). It is recommended that roguing should be carried out in the early morning hours before pollen shedding takes place. The R-lines should also be rogued periodically.

Hybrid seed harvested from the male sterile plant (A-line) is the main interest to producers and thus maximum percentage seed set on the male sterile line is the most important objective. Usually, the R-line is harvested first, the A-line rows are carefully inspected for off-types and other admixtures and then harvested. Seed of the R-line harvest is generally not to be reused as seed; for hybrid (A × R) seed production again in the next season.

Seed standard and Field standards	F/S	C/S
Pure seed (min.)	98%	98%
Inert matter (max.)	2%	2%
Other crop seed (max.)	2/kg	2/kg
Weed seeds (max.)	5/kg	10/kg
Germination (mini.)	75%	75%
Moisture (max.)	12%	12%
Pollen shedders (max.)	0.05	0.10
Off-types (max.)	0.05	0.10
Diseased head (max.)	0.05	0.10
ODV	10/kg	20/kg

Synchronization of Flowering

As two varieties differing in flowering dates are involved in hybrid seed production, its success mainly depends upon, how best the two varieties are manipulated to flower simultaneously to get highest seed setting.

1. Sowing of female and male parent on different dates.
2. In case of difference, giving the additional dose of nitrogen in the soil followed by foliar application to early flowering line.
3. If male is advanced, cut alternate plants to allow the tillers to come up.
4. Spraying growth retardant Maleic hydrazide 500 ppm at 45th day after sowing delays flowering in advancing parent.
5. Nitrogen in the form of urea 1% can be sprayed to the lagging parent.
6. With held one irrigation in advancing parent.
7. Take up staggered sowing.
8. Spraying CCC 300 ppm will delay flowering.

Control of Pest and Diseases

Use of prescribed fungicides/ insecticides for the control of insect-pests and diseases. Apply 40kg/hect. Phorate granules or 20 kg/hect. Carbofuran granules in furrows before sowing to control shoot fly. For control of stem borer and foliar diseases spray the crop with 36 ml Endosulfan and 40 g zineb in 18 lit. of water, 25 and 45 days after sowing.

For control of Sugary disease, spray 40g Ziram or 45g Thiram in 18 lit. of water before the emergence of boot leaf at 50% panicle initiation and also at flowering.

In order to control Ear head midge, Bug and Ear head Caterpillar, spray the crop at 50% panicle emergence stage with 36 g Carbaryl or 35 ml Endosulfan or 36 ml Malathion in 18 lit. of water. Repeat the spray after 4 or 5 days.

Hybrid Pearl Millet

Pearl millet is commonly known as Bajra, Sajje, Cumbo, Cattail, millet, Bull rush millet, Candle millet or Dark millet. It belongs to family Poaceae and has two important species (i) *Pennisetum glaucum* (2n = 14) grown for grain purpose and (ii) *Pennisetum perpereum* (2n = 56) grown for fodder grass purpose.

Pearl millet is highly protogynous and predominantly a highly cross pollinated (80%) though self-pollination to an extent of 20-30 per cent has been reported in this crop. Complete emergence of all styles takes nearly 2-3 days after panicle exsertion from boot leaf. The stigma is receptive for 12–24 hours after its emergence and is extended till a day after exsertion of anthers. Anthesis is completed in two phases. Anthesis in bisexual flowers occurs first followed by Staminate flowers. Generally the anthesis in sessile male florets emerge 2–3 days after the anthesis of bisexual flowers. The largest number of anthers emerge between 8 AM – 2 PM reaching a peak at 10 AM. The anthesis starts from apex to base. Pollen shedding in a head may be continued upto 4–6 days. Pollens are viable only for two hours.

After the discovery of CGMS (line (Tift 23A) in 1958 at Tifton, Georgia (USA) by Dr. G.W. Burton, the hybrids were developed in India. During 1965 a first bajra hybrid HB–1 was developed in Ludhiana by Dr. D.S. Athwal utilizing MS 23A × Bill 3B.

Hybrid Seed Production Techniques

(*i*)   Maintenance of parental lines A, B and R.

(*ii*)   Production of hybrid seed (A × R).

Land Requirements

Land to be used for seed production should be free from volunteer plants. If bajra was grown in the previous season, irrigate the land 3 days before sowing and plough just before sowing so as to eliminate the volunteer bajra plants. Land should be uniform, leveled with good drainage.

Isolation	F/S	C/S
Field of other varieties including commercial hybrid of the same variety	1000 m	200 m

(Contd...)

Table Contd...

Isolation	F/S	C/S
Field of same hybrid not conforming to varietal purity requirements	1000 m	200 m
Field of other hybrids having common male parent but not conforming to varietal purity requirements	—	200 m
Field of other hybrids having common parent and conforming to varietal purity requirement	—	5 m

Planting Ratio

The recommended planting ratio of female to male rows is 4 : 2 with 8 border rows of male parents. The male rows are to be marked for identification and to reduce the chances of errors in planting.

Spacing

Line to line	75–90 cm
Plant to plant	15–20 cm

Seed Rate

2.5 kg. Female and 1.5 kg. male per hectare for hybrid seed production (A × R) and 4 kg/hect. for B and R-line multiplication.

Seed Treatment

Seed treated with Apron 35 SD @ 4g/kg. to check downy mildew infection.

Fertility

100 : 60 : 40 N, P_2O_5 and K_2O kg/hect. Apply 40 kg N and full dose of phosphorus and potash before planting as basal and the remaining 60 kg N as top dress after 25–30 days of sowing.

Synchronization

Synchronization of male and female parent is essential for obtaining higher seed yield and seed setting. Emerging panicles of advanced lines are pulled out (jerked) to facilitate tiller growth. Urea spray to late parent enhances growth and hastens flowering. However, if the difference in flowering is more than 10–15 days then staggered sowing of early flowering parent is needed.

Roguing

Rogue out the volunteer plants, off-types, diseased plants, pollen shedders etc., before flowering by cutting such plants close to ground or by uprooting, so as to prevent regrowth.

Field Inspection

Four field inspections are carried out as follows :

First : It is done before flowering to determine isolation, presence of off-types and planting ratio.

Second and Third : This is done during flowering to determine pollen shedding in female plants, off-types plants and diseased plants etc.

Fourth : It is made prior to harvest when the seed is mature enough to reveal its true to type and also to verify disease levels.

Harvesting

Male rows are harvested before the female rows to avoid mechanical mixture.

Field Standards

	F/S	C/S
Off-types	0.05%	0.10%
Pollen shedder	0.05%	0.10%
Plants affected with downy mildew/green ear/green smut	0.05%	0.10%
Ergot ears	0.02%	0.04%

Seed Standards

Pure seed (mini.)	98%	98%
Inert matter (max.)	2%	2%
Other crop seed (max.)	10/kg	20/kg
Weed seed (max.)	10/kg	20/kg
Ergot (max.)	0.02	0.045
Germination (mini.)	75%	75%
Moisture	12%	12%

Oat (*Avena sativa* L.)

Oat is a cereal forage crop of family Poaceae. It is commonly known as Jai, and it is a widely cultivated forage crop in India.

In Oat, the inflorescence is compound, comprising a series of flowering branches and spiklets. The inflorescence terminates the stem in the form of a panicle. The reproductive organs consist primarily of three stamens, their filaments and anthers, the single ovary with its style and the bifid stigma.

Oat is a self-pollinated crop. Natural cross pollination by wind occurs occasionally, the extent of which varies from 0.40–1.3 per cent.

Land Requirement

The field selected for Oat seed production should not have Oat crop in previous year. The seed plot should be free from volunteer plants.

Isolation

An isolation of at least 3 meters around the seed field from the other fields of the Oats is necessary for foundation and certified seed production.

Seed Rate – 75 to 100 kg/hect.

Spacing – Line to line – 23 cm.

Fertilizers

Usually 80–100 kg nitrogen and 60 kg phosphorus per hectare is required. Half dose of nitrogen and entire dose of phosphorus should be applied before sowing of the crop. The remaining half quantity of nitrogen is applied in two split doses.

Irrigation

Irrigation should be given after gemination and at 3 weeks intervals depending upon the local conditions.

Roguing

Roguing of seed production plots is done to remove the off-type plants that arise due to segregation of residual heterozygosity, out crossing with other varieties, admixture or mutations etc. A minimum of three roguings must be done. First roguing should be done just ahead of the flowering stage. At this stage, off-type plants, smutted plants and early heading plants should be removed. Subsequent roguing should be done just after flowering is complete and the panicles start to turn colour. At this stage, tall plants and late plants can be identified. Roguing at the stage of panicle acquiring colour, identifies the plants which are off-types at reproductive stage.

Inter-Culture and Weeding

One weeding after two or three weeks is necessary to control the weeds. The broad leaf dicot weeds can also be controlled by applying selective herbicide, such as 2–4–D. The 2–4–D should be sprayed when the plants attain 15 to 20 cm height.

Harvesting and Threshing

The harvesting should be done as soon as the crop matures. Any delay in harvesting may result in deterioration of seed quality. The crop can be harvested manually and threshed in such a way so that the mechanical admixture may not be arise.

Seed Standards

Pure seed (min.)	98%	98%
Inert matter (max.)	2%	2%
Other crop seeds (max.)	10/kg	20/kg
Total weed Seeds (max.)	10/kg	20/kg
Objectionable weeds (max.)	2/kg	5/kg
O.D.V. (max.)	10/kg	20/kg
Germination (min.)	85%	85%
Moisture (max.)	12%	12%

5
SEED PRODUCTION OF PULSES

CHICKPEA (*Cicer arietinum*)

Chickpea is predominantly a self-pollinated crop and extent of out crossing remain below 5%. The flowers being cleistogamous, typical of subfamily Papilionoideae, the reproductive part remain enclosed in the keel. The white bud stage, when petals emerge from the calyx but are still enfolded, is the most crucial reproductive phase. At this stage, the anther's filament elongate and keep the anthers close to stigma at the

Important Diagnostic Characteristics

Character	Status	Stage of observation
Plant type	Erect/Semi erect/Semi spreading/prostrate	Vegetative
Days to flowering	Early/medium/late	Flowering
Flower colour	White/pink/blue	Flowering
No. of flowers and pods/peduncle	Single/twin/multiple	Reproductive
Pod Size	Short/medium/long	Maturity
No. of seeds/pod	Single/double/triple	Maturity
Seed shape	Pea shaped/owl's head/angular	Maturity
Testa texture	Rough/smooth/tuberculated	Maturity
Seed Colour	Beige/brown beige/green/yellow/brown/dark brown/pink	Maturity
Seed Size	*Desi* : Bold/medium/small *Kabuli* : Extra bold/bold	Maturity

time of dehiscence. Full bloom occurs approximately 24 hrs. after pollen is shed and thus leaving a very little chance of cross-pollination.

Land Requirement

The land selected should not have been under the same crop in previous season.

Isolation

Since the crop is highly self-pollinated, an isolation of 10 meters and 5 meters will be required for F/S and C/S, respectively.

Spacing

Row to row – 30 to 45 cm.

Seed Rate

75 – 80 kg/hectare (small seeded).

90 – 100 kg/hectare (bold seeded).

Fertilizers

A basal dose of 25 kg nitrogen and 50 kg of phosphorus is applied.

Roguing

All the off-type plants are to be rogued out at vegetative, flowering and at maturity stage.

Harvesting and Threshing

Harvest the crop manually and threshed with care that mechanical admixture can not be in seeds.

Field Standards	F/S	C/S
Off-types plants (max.)	0.10%	0.20%
Objectionable weed plants (max.)	–	–

Seed Standards	F/S	C/S
Pure seed (mini.)	98%	98%
Inert matter (max.)	2%	2%
Other crops seed (max.)	2/kg	5/kg
Total weed seeds	–	–
Objectionable weed seeds	–	–
O.D.V. (max.)	5/kg	10/kg
Germination (mini.)	85%	85%
Moisture (max.)	9%	9%

LENTIL (*Lens culinaris*)

Lentil is a self-pollinated crop and the extent of out crossing is not above 5%.

Land Requirement

Loam and clay loam soils are suited for its cultivation. The field which is selected for seed production free from volun-teer plants. In the previous year there should not be lentil crop.

Isolation Distance

An isolation distance of 10 meters and 5 meters should be kept for foundation and certified seed production, respectively.

Sowing Time

Lentil crop should be sown in mid October to mid November for quality seed production.

Seed Rate

For timely sowing 40–60 kg/ha. seed is required. The seed should be treated with Rhizobium culture @ 200 g/10 kg seed.

Spacing

Line to line : 25 cm

Plant to plant : 10 cm

Fertilizers

Apply 20 kg nitrogen, 60 kg phosphorus and 20 kg potash per hectare and 20 kg/ha. sulphur before sowing.

Irrigation

One irrigation may be provided before flowering.

Interculture

Two hand weedings at 30 and 60 days after sowing of the crop. For weed control application of pre-emergence weedicide *i.e.,* Pendimethalin @ 1.5 kg/ha.

Plant Protection

The important insect-pests of Lentil is Aphid and for its control one spray of Cypermethrin 0.004%, Dimethoate 0.03%, Endosulphan 0.07% or Malathion 0.05% should be done.

The important diseases of Lentil are wilt and rust. For the control of wilt seed must be treated with Benomyl + Thiram (1 : 1) @ 3g/kg seed before sowing. For rust seed dressing with Agrosan GN @ 2.5g/kg. seed or spray Zineb @ 2.5 g per lit. water.

Roguing

Off-type plants and diseased plants must be rogued out from the seed plot.

Harvesting and Threshing

Harvesting of crop should be done after maturity. Care should be taken during harvesting and threshing to avoid mechanical admixture.

Field Standards	F/S	C/S
Off-types plants (max.)	0.10%	0.20%

Seed Standards	F/S	C/S
Pure seed (min.)	98%	98%
Inert matter (max.)	2%	2%
Other crop seeds (max.)	5/kg	10/kg
Total weed seeds (max.)	10/kg	20/kg
Objectionable weed seeds (max.)	–	–
O.D.V. (max.)	10/kg	20/kg
Germination (min.)	75%	75%
Moisture (max.)	9%	9%

FIELD PEA (*Pisum sativum*)

Field pea is a pre-dominantly self-pollinated crop and the extent of out crossing is not above 5%. The flowers are cleistogamus and reproductive parts remain enclosed in the keel which is a typical characteristic of sub-family Papilio-noideae. The most crucial reproductive phase is the stage when petals come out from the calyx. At this stage, filament of anthers elongate and put the anther near to stigma for dehiscence.

Land Requirement

Loam and sandy loam soils are well suited for pea seed production. In previous year where pea crop was taken should not be considered for seed production of field pea. The selected field should be free from volunteer plants.

Isolation

For foundation and certified seed production an isolation distance of 10 meters and 5 meters, respectively should be provided.

Seed, Season and Sowing

Sowing of field pea may be done in rows from middle of October to middle of November.

Seed Rate

80 – 100 kg/ha. tall type

100 kg/ha. dwarf type

Spacing

For tall type varieties the spacing should be kept at 30 × 10 cm and for dwarf varieties the spacing may be 22.5 × 10 cm.

Fertilizers

For tall varieties apply 20 kg nitrogen, 40 kg phosphorus and 20 kg K_2O per hectare. Apart from these fertilizers apply sulphur @ 20 kg/ha. for seed production. However, for dwarf varieties the quantity of fertilizers is 40 kg nitrogen, 40–60 kg phosphorus, 20 kg potash and 20 kg sulphur/ha.

Irrigations

For field pea the requirement of irrigations may be 2-3.

Interculture

Two hand weedings at 30 and 60 days after sowing. There should be pre-emergence application of Pendimethalin @ 1.5 kg/ha.

Plant Protection

For the control of stem fly or leaf minor seed must be treated with 2% Phorate before sowing or soil treatment with Phorate at the time of sowing @ 1 kg/ a.i/hectare. For control of powdery mildew spray wettable sulphur @ 3g/lit or Dinocap @ 2g/lit water and for rust spray with Maneb @ 2g/lit. water.

Harvesting and Threshing

After maturity the crop should be harvested and threshed without any mechanical admixture. After proper cleaning and drying the seed should be stored.

Field Standards	F/S	C/S
Off-types plants (max.)	0.10%	0.20%

Seed Standards	F/S	C/S
Pure seed (min.)	98%	98%
Inert matter (max.)	2%	2%
Other crop seeds (max.)	2/kg	5/kg
O.D.V. (max.)	5/kg	10/kg
Germination (min.)	75%	75%
Moisture (max.)	9%	9%

GREEN GRAM (Mung Bean)

Mung bean is strictly self-pollinated crop and the extent of out crossing is not above 5%. The flowers are cleistogamus and reproductive parts remain enclosed in the keel which is a typical characteristic of sub family papilionoideae.

Land Requirement

Mung bean can be cultivated successfully in light and well drained loam as well as red and black soils. Seed field should not have been cropped with same crop in the previous season. The field should be free from volunteer plants.

Isolation

Foundation seed : 10 meters

Certified seed : 5 meters

Spacing

Line to Line : 30 to 35 cm

Plant to Plant : 7 to 10 cm

Seed Rate

For sowing of 1 ha. area 15–20 kg seed is required. The seed must be treated with Thiram @ 2 g/kg. seed. The seeds should be treated with Rhizobium culture @ 10 kg/ha.

Fertilizer Application

The recommended dose of 15 kg nitrogen, 40 kg phosphorus and 20 kg sulphur per hectare may be applied before sowing. The fertilizer should be placed 7.5 cm away from the seed row.

Irrigation and Interculture

For kharif crop, one or two irrigations may be required if the rains fail and the dry period is prolonged. In case of summer crop frequent irrigations are required.

To suppress the growth of weeds, one or two weedings followed by one or two hoeings are needed.

Plant Protection

For the control of pod borer spray the crop with Endosulphon 35 Ec @ 1.25 lit/ha. or Quinalophos 25 EC @ 1.25 lit ha.

For the control of yellow mosaic spraying of Dimethoate 30 EC @ 1 lit (600–800 lit of water)/ha. or Methyl-o-Dematon 25 EC. @ 1 lit/ha. (600–800 lit water) will be required.

Roguing

All the Off-types and diseased plants should be rogued out as and when they are identified in the seed plot.

Harvesting and Threshing

The crop should be harvested by picking the pods after they have turned black after maturity. After drying the pods threshing should be performed in such a manner that mechanical admixture may not be happen in seeds.

Field Standards	F/S	C/S
Off-types plants (max.)	0.10%	0.20%
Plants infected with disease (max.)	0.10%	0.20%

Seed Standards	F/S	C/S
Pure seed (min.)	98%	98%
Inert matter (max.)	2%	2%
Other crop seeds (max.)	5/kg	10/kg
Total weed seeds (max.)	5/kg	10/kg
Germination (min.)	75%	75%
Off-type (max.)	0.10%	0.20%
O.D.V. (max.)	10/kg	20/kg
Moisture (max.)	9%	9%

URD BEAN

Urd is a predominantly self pollinated crop and the extent of out crossing is not above 5%.

Land Requirement

The land should be levelled and well drained. It can be cultivated successfully in light loam as well as red and black soils. Seed field should not have been cropped with Urd bean in the previous season.

Isolation

An isolation distance of 10 meters and 5 meters should be kept in foundation and certified seed production.

Spacing

Spring/Summer	Kharif
Line to line : 30 to 35 cm	45 cm
Plant to plant : 7 to 10 cm	10 cm

Seed Rate

For summer crop the requirement of seed for sowing of 1 hectare area is 30–35 kg. However, for Kharif crop the seed rate should be 20–25 kg./ha.

Planting Time

In summer/spring season the crop should be sown in 2nd fortnight of March to first week of April. For Kharif crop the sowing may be completed at the onset of monsoon.

Irrigation and Interculture

For summer crop first irrigation may be provided at 25 days after sowing. Other irrigations may be provided at 7–10 days interval as per requirement.

Fertilizers

For summer crop apply 10 kg nitrogen, 30 kg phosphorus, 20 kg potash and 20 kg sulphur/hectare and for Kharif crop the requirement of fertilizers is 20 kg nitrogen, 40 kg phosphorus, 20 kg potash and 20 kg sulphur per hectare.

Roguing

All the off-types and diseased plants should be rogued out from the seed field time to time.

Weed Management

One hand weeding at 20 days after sowing should be done. There must be pre-emergence application of Pendimethalin @ 1.5 kg/ha. for the control of weeds in Urd crop.

Plant Protection

Yellow mosaic virus, Cercospora leaf spot and powdery mildew are the important diseases of Urd bean. For control of yellow mosaic virus, spray the crop with Malathion or Meta-systox @ 1 ml/lit. water. For Cercospora leaf spot Carbendazim @ 0.025% may be sprayed at 30 and 45 days after sowing. Spraying of Urd crop with wettable sulphur @ 3g/lit. or Dinocap @ 1 ml./lit. water should be done for the control of powdery mildew.

For the control of insects and pests in Urd there may be soil application with phorate or Carbofuran granules @ 1 kg/

a.i./ha. before sowing of summer crop. However, in Kharif crop against stem fly, jassid and white fly, two additional sprays with Dimethoate @ 0.03% or Monocrotophos 0.04% at 45 and 60 days after sowing against foliage feeders and pod borers.

Harvesting and Threshing

Matured pods can be plucked manually or at maturity harvest the crop and after drying threshing should be performed without any mechanical admixture. In spring/summer crop care must be taken that before the onset of monsoon all the matured pods should be plucked. After proper drying the pods should be threshed manually or by bullock treading.

Field Standards	*F/S*	*C/S*
Off-type plants (max.)	0.10%	0.20%

Seed Standards	*F/S*	*C/S*
Pure seed (min.)	98%	98%
Inert matter (max.)	2%	2%
Other crop seeds (max.)	5/kg	10/kg
Total weed seeds (max.)	5/kg	10/kg
O.D.V. (max.)	10/kg	20/kg
Germination (min.)	75%	75%
Moisture (max.)	9%	9%

RAJMASH

Rajmash is a predominantly self-pollinated crop with 5% out crossing. It is generally grown in rabi season.

Land Requirement

During rabi season it can be grown in loam and sandy loam soils with well drained facility. The seed production plot should be free from volunteer plants.

Isolation

An isolation distance of 10 meters and 5 meters is desirable for foundation and certified seed production.

Seed Rate

For planting of 1 hectare area the requirement of seed will be 120–140 kg.

Spacing

Line to Line : 30 to 40 cm

Plant to Plant : 10 cm

Sowing

After seed treatment with Thirum @ 2.5g/kg. seed, sowing may be done from 3rd and 4th week of October to first week of November.

Fertilizers

Apply 90–100 kg nitrogen, 60 kg phosphorus, 30 kg potash and 20 kg sulphur/ha. Half dose of nitrogen and entire quantity of phosphorus, potash and sulphur should be applied as basal dose. Remaining quantity of nitrogen must be given as top dressing in the crop. Spray of 2% urea solution at 30 and 50 days after sowing the seed yield may be increased.

Roguing

All the Off-types and diseased plants should be removed from the seed plot before flowering and also when observed.

Interculture

Hoeing and weeding must be done just after first irrigation. At the time of hoeing there may be earthing up the plants which protected from lodging during fruiting.

Irrigations

Two to three irrigations are required in Rajmash. The first irrigation is given at 28–30 days after sowing. Rest irrigations may be given at one month interval. During irrigation there should not be stagnation of water in the field.

Plant Protection

For the control of mosaic spray the crop with Rogor/ Dimecron @ 1.5 cc/lit. to water. Mosaic affected plants must be removed from the seed plot immediately.

Harvesting and Threshing

After maturity of the pods the crop should be harvested. After drying the crop must be threshed and ensured no mechanical contamination.

Field Standards	F/S	C/S
Off-type plants (max.)	0.10%	0.20%
Diseased plants (max.)	0.10%	0.20%

Seed Standards	F/S	C/S
Pure seed (min.)	98%	98%
Inert matter (max.)	2%	2%
Total weed seeds (max.)	–	10/kg
O.D.V. (max.)	5/kg	10/kg
Germination (min.)	75%	75%
Moisture (max.)	9%	9%

COWPEA (*Vigna sinensis*)

It is most common crop grown in the northern than in the southern plains. It is primarily grown for green vegetables like other beans.

Land Requirements

The field selected for seed production should be free from volunteer plants. The land should have light type soil with good drainage.

Isolation Requirement

The isolation should be 10 meters in case of foundation and 5 meters in case of certified seed.

Season, Seed and Sowing

The seed can be sown as dibbling with row to row spacing of 45–60 cm and plant to plant distance should be 10–15 cm. The crop is grown during Kharif and Summer. During Kharif it can be sown in June-July or August-September and during summer it can be sown in February to March.

Seed Rate

25–30 kg/ha.

The seed should be treated with Rhizobium culture @ 375 g/hectare.

Fertilizers

Apply 25 kg nitrogen, 50 kg phosphorus and 25 kg potash per hectare. Nitrogen can also be supplied in split doses.

Irrigation and Interculture

In case of summer crop regular irrigations has to be provided till the crop is matured. One weeding and two hoeings are needed to keep the seed field free from weeds before flowering.

Plant Protection

The major pests of the crop are pod borer, aphid and white fly. The main diseases are bacterial blight, anthracnose and cow pea mosaic.

The seed should be treated with Captan or Thiram @ 2g/kg seed. When the pests are observed spray with 30 ml of Dimethioate or 9 ml of Phosphomedon in 18 lit. of water. About 500 lit. of solution is required for spray of 1 hect. area. Second spraying may be done after one week if infection persist.

In case of disease, the crop should be sprayed with 50g soluble sulphur or 4g Zineb or Mancozeb or 60 g copper oxychloride dissolve in 18 lit. of water. If the crop is infected with bacterial blight spray the crop with Agromycin @ 9 g dissolved in 18 lit. of water.

Roguing

All the Off-types and diseased plants should be removed from seed plot as and when observed.

Harvesting and Threshing

When the pods were matured and dry harvesting can be done manually with the help of sickle. After drying threshing should be performed. Care should be taken during harvesting and threshing to avoid mechanical admixture.

Field Standards	F/S	C/S
Off-type plants (max.)	0.10%	0.20%
Diseased plants (max.)	0.10%	0.20%

Seed Standards	F/S	C/S
Pure seed (min.)	98%	98%
Inert matter (max.)	2%	2%
Other crop seeds (max.)	–	10/kg
Total weed seeds (max.)	–	10/kg
O.D.V. (max.)	5/kg	10/kg
Germination (min.)	75%	75%
Moisture (max.)	9%	9%

HYBRID PIGEONPEA

Pigeonpea is commonly called as Arhar, Tur, Red Gram, Angola pea and Congobean. Botanically it is called as *Cajanus cajan* (L.) Millspaugh, with chromosome number, 2n = 22. It belongs to family, Fabaceae (Leguminosae), subfamily, Papilonaceae.

In pigonpea, inflorescence is produced as axillary receives with 3–12 flowers and flowering proceeds acropetally. Flowering is basipetal in the determinate cultivars. The yellow anthers (10) surround the stigma. Dehiscence of anthers takes place at least 24 hours before the flower opens, hence self-pollination occurs. Occasional out crossing is caused by thrips and bees. It has a higher natural crossing rate than any other pulse crop. The extent of outcrossing varies with locations and the insect species involved. The outcrossing percentage varies from 3% to as high as 40% depending on the locations, the average is around 20% (Bhatia *et. al.,* 1981). The stigma is receptive 68 hours before anthesis and continues for 20 hours after the anthesis. In pigeonpea, F_1 hybrid seed can be produced by using genetic male sterility system in two main stages :

Maintenance and Multiplication of Male Sterile (GMS) Line

The GMS line is maintained by crossing the homozygous recessive genetic male sterile parent with dominant heterozygous male fertile line in 1 : 1 planting ratio in an isolated field. Such a crossing gives rise to a progeny in ratio of 1 : 1 male sterile and fertile plants. Simultaneously in another isolated field, the dominant homozygous male fertile line is also raised as a single pure crop. The resultant seed is used for future production of F_1 hybrid seed. The production of seeds of MS line and pollinator line is referred to as foundation seed production.

Production of F_1 Hybrid Seed

F_1 seed is produced by crossing the genetic male sterile line with homozygus dominant male fertile line in the prescribed planting ratio in an isolated field. The resultant seed is harvested as F_1 hybrid seed.

The entire progeny of the first cross is used as female parent and planted with homozygous dominant male fertile line. In the female rows, both sterile and fertile plants are present. The fertile plants should be removed from female rows before they shed their pollens. At flower initiation stage, fertile plants are examined by two methods *i.e.,* (*a*) physical observance for presence of fertile or sterile pollens (*b*) seedling gene markers.

The plants with fertile anther's should be removed immediately and this roguing should be continued daily until last of the plant of the female rows is completed. The remaining plants are sterile plants and are allowed to be cross pollinated with dominant pollinator line to get F_1 hybrid seed. This stage of seed production is called as certified hybrid seed production.

Land Requirements

Pigeonpea crop can be grown over a wide variety of well drained soils having neutral pH (5–7). It can be grown in sandy loam, alluvial and deep black soils. It can not tolerate excessive acidity (pH < 5.0). The land should be well fertile. It should not be grown with the same crop in the previous season to avoid occurrence of volunteer plants and seed borne diseases.

Isolation

It is a mainly self-pollinated crop but it exhibits high natural crossing of about 60–80 per cent depending upon the locations and insect species involved. The minimum isolation distance is 200 m and 100 m for foundation and certified seed production, respectively.

Sowing Time

June–July is the best time for hybrid seed production of pigeon pea.

Seed Source

B/S or F/S seeds should be obtained from authentic and approved source.

Planting Ratio

For hybrid seed production a planting ratio of 4 : 2 or 6 : 2 or 4 : 1 or 6 : 1 male sterile to pollen parent is to be adopted depending upon the honey bee activity in the area.

Spacing

Row to row : 60 to 75 cm

Plant to plant : 25 to 30 cm

Fertilizers Requirement

25 kg nitrogen and 50 kg P_2O_5 per hectare is recommended. Apply ½ dose of nitrogen and full dose of phosphatic fertilizers at time of sowing. Remaining half dose of nitrogen is to be given at 30-35 days after sowing. The fertilizer is to be placed at 10–15 cm deep into the soil and also at side of the seeds.

Method of Sowing

The designated land should be laid out with prescribed planting ratio of female and male rows. Border rows are normally provided. The male rows should be marked at both ends by wooden pegs or marker plants. In the respective parental rows, seeds should be sown manually by dibbling or using a seed drill. Both parents are to be sown on the same days. The seed should be sown at a depth of 3–5 cm at adequate soil moisture conditions (at least 40–50% moisture availability).

Seed Treatment

Seeds are treated with specific rhizobium strains @ 600 g/hectare to get better nodulation and organic fertility in the soil. Phosphobacteria @ 600 g/hectare as seed treatment is also recommended for increasing the phosphorus use efficiency.

In order to withstand drought and to get early germination and vigour with available soil moisture pigeonpea seeds should be treated with trichoderma @ 4 g/kg followed by rhizobium culture nodulation at 24 hours interval and subsequently pelleted with $ZnSO_4$ @ 100 mg/kg and Gypsum @ 300g/kg as carrier and maida 10% (in 50 ml) as adhesive.

Interculture and Weed Management

Pigeonpea is slow growing during the first 40–50 days of its growth. It is very sensitive to weeds during early growth stage. At least one or two hand weeding and hoeings are to be followed in early stages of growth.

Pre-emergent application of Lasso (4 lit/hectare) or Penda-methaline can also be used to control weed population in seed plots.

Plant Protection

Highest losses are due to pod borer (*Heliothis armigera*) causing 12.6 – 34.1% damage while, pod fly (*Melangomyza obtusa*) can cause 9.0–24.8% damage.

Diseases

Wilt (*Fusarium udum*), phytophthora blight (*Phytophthora dreschlori* f.spi.cajani), collar rot (*Sclerotium rolfsii*), dry root rot (*Rhizoctonia bataticola*) and leaf spot (*Cercospora cajani*), sterility mosaic, yellow mosaic virus and witches broom (mycoplasma) are the major diseases in pigeonpea.

Control Measures

Wilted plants should be removed as and when found in the seed field. Spray the crop with Endosulfon 25 EC or Quinalphos 25 EC @ 36 ml or 18 ml Monocrotophos 40 EC mixed with 18 lit. of water. The spray can be repeated after 15 days if the infection persist. To avoid pulse bettle infestation in storage, the crop should be sprayed with Endosulfan (0.25%) or Malathion (0.07%) either singly or in combination with Carbendazim (0.1%) two times at weekly interval before harvest of the crop.

Roguing

To maintain the desired genetic purity and physical purity of hybrid seed, roguing should be done timely from vegetative to harvesting stage. The Off-types and volunteer plants are in the field based on leaf colour, size, stem colour, growth habit, flower colour, pod colour, seed colour etc.

The pest and diseases affected plants should be removed from the seed plot.

In the male sterile rows (female) particularly during flowering, roguing should be done as following :

(*i*)   Remove the Off-types and diseased plants.

(*ii*) Remove the male fertile plants by observing the anther colour (yellow) at the time of first flower formation. The plants with translucent white anthers (sterile) alone should be retained in the female rows. This removal should be continued daily for at least 7–10 days until the completion of flowering. Since the female parent is genetically male sterile, 50% of the male fertile plant population should be rogued out from female rows before they shed their pollens.

(*iii*) Also, remove the late flowering/maturing and early flowering/maturing plants.

In the Pollen Parent (Male Fertile Rows)

(*i*)   Remove all Off-types and diseased plants.

(*ii*) Remove the immature pods set early in the plants timely to induce the continues flowering and make pollen availability for longer period.

Harvesting

The pigeonpea hybrid seed should be harvested from the female plants when the seeds are fully matured (pods turned straw colour). The male row plants are harvested first and kept away from the main seed field. Every care is to be taken to avoid mechanical admixtures during harvesting and threshing.

Field Inspections

A minimum of two field inspections shall be made in such a way that at least one of them is made during flowering.

Field Standards	F/S	C/S
Off-types at and after flowering	0.10%	0.20%
Plants affected by seed borne diseases	0.10%	0.20%

Seed Standards	F/S	C/S
Pure seed (min.)	98%	98%
Inert matter (max.)	2%	2%
Other crop seeds (max.)	5/kg	10/kg
Weed seeds (max.)	5/kg	10/kg
O.D.V. (max.)	10/kg	20/kg
Germination (min.)	75%	75%
Moisture (max.)	9%	9%
Genetic purity (mini.)	99%	95%

CLUSTER BEAN
(*Cyamopsis tetragonoloba* L. Taub)

Cluster bean or guar is one of the most hardy legume vegetable and successfully cultivated in North-West and Southern parts of India where rainfall is low. The green pods are consumed as vegetable. Seeds contain gum like mucilaginous substance called guar gum or galactomannan which is used in textiles, paper industry and cosmetic industry.

It is a warm season crop and grows well in both summer and rainy seasons in the plains. It is a hardy plant and tolerant to drought conditions.

Land Requirement

It can be grown on wide range of soils with adequate facility of drainage. The field which is used for seed production should be free from volunteer plants.

Fertilizers

Cluster bean is a leguminous crop. It requires less amount of nitrogen. Apply 10–20 kg nitrogen, 60–70 kg phosphorus and 70–80 kg potash per hectare. Entire dose of fertilizers should be applied as basal dressing.

Seed Rate

Seed rate varies from 20–40 kg/ha.

Sowing

In southern plains crop is sown during December–January. In plains it is grown twice in a year *i.e.,* February to March and June–July.

Spacings

The seeds are sown 2–3 cm deep in rows kept 45–60 cm apart. The distance between plants to plants is maintained 15–30 cm.

Irrigations

Irrigate the crop as and when required.

Interculture and Weed Control

About 2–3 shallow hoeings are required to keep the field free from weeds. Herbicides like Fluchloralin @ 1.0 kg/ha as pre-plant emergence and Alachlor @ 1.5 kg/ha or Nitrogen @ 3.0 kg/ha. or Pendimethalin @ 2.0 kg/ha. as pre-emergence can also be used to control the weeds effectively.

Roguing

All the Off-types and diseased plants should be rogued out from seed plot as and when they observed.

Harvesting and Threshing

When the pods are matured picked and dried. After proper drying the threshing is done manually by stick beating and cleaned properly.

Plant Protection Measures

Bacterial blight is the most serious disease of cluster bean. This disease is seed and soil borne both. Treat the seeds with Carbendazim @ 3 g/kg seed or Thiram @ 2.5 g/kg seed.

For the control of blight soak the seeds in a mixture of Streptocycline (1g) and Hexacap (2.50g) in 10 lit. of water for 4 hours before sowing.

Seed Standards

	F/S	C/S
Pure seed (min.)	98%	98%
Inert matter (max.)	2%	2%
Other crop seeds (max.)	10/kg	20/kg
O.D.V. (max.)	10/kg	20/kg
Germination (min.)	70%	70%
Moisture (max.)	9%	9%

6

SEED PRODUCTION OF OIL SEED CROPS

RAPE SEED – MUSTARD

Rape seed-mustard is a unique group of crops having different kinds of breeding behaviour. On the one hand are self-compatible (self-pollinated) crops like yellow sarson *Brassica campestris* var. yellow sarson, Gobhi sarson (*Brassica napus*) and Ethiopian mustard (*Brassica carinata*) while on the other hand are self-incompatible (cross-pollinated) crops including Toria (*Brassic rapa* var. toria), brown sarson (*Brassica rapa* var. brown sarson) and taramira (*Eruca sativa*).

In Indian mustard (*Brassica juncea* L. Czern and Coss), the major crop of the group, cross-pollination ranges from 5 to 15 per cent.

Important Diagnostic Characters

Indian Mustard : Leaves are dark green, syrate, pinnatified petiolated.

Rape Seed : Leaves sessile.

Land Requirement

It should be ensured that the selected field should not have been under *Brassica* spp. in the previous season. The field should be free from volunteer plants.

Isolation

	F/S	C/S
For self-compatible	50 m	25 m
For self-incompatible	100 m	50

Spacing

Line to line : 45 cm

Plant to plant : 10–15 cm

Seed Requirement

4 kg/ha. toria

5 kg/ha. mustard and yellow sarson

Fertilizers

For toria 80–100 kg nitrogen, 50 kg Phosphorus and 50 kg Potash/ha. and for Indian mustard 120 kg Nitrogen, 60 kg Phosphorus and 60 kg Potash/ha. and 40 kg Sulphur per hectare should be applied. Full dose of phosphorus, potash, sulphur and half quantity of nitrogen should be applied as basal dose and the remaining half quantity of nitrogen should be top dressed after first irrigation (25–30 days after sowing).

Thinning

After 15–20 days of sowing thinning must be done and plant to plant distance is maintained approximately 10–15 cm.

Plant Protection

1. **Alternaria blight :** Alternaria blight appears on the leaves of one month old crop in the form of brown or black spots with concentric rings. The disease may be controlled by spraying of Iprodione or Mancozeb @ 0.2%, Ziram (27%) @ 3.5 lit/ha., Copper oxychloride (80%) @ 3 kg/ha.

2. **White Rust :** It appears on the lower side of the leaves as white, raised, dry, eruptions. Seed treatment by Apron 35 SD @ 6 g/kg. seed is recommended to prevent the disease occurrence. To control the disease spray of Ridomil M.Z. @ 0.25% at 50 days after sowing is recommended.

3. **Sclerotinia Stem Rot :** It is a soil borne disease emerging as an important disease in recent years. Its symptoms appear as water soaked pale or brown lesions at base of stem. The black pellet–like sclerotinia may be formed either internally in the pith of the stem or on the stem surface. The seed should be treated with 0.1% a.i. Carbendazim. Foliar spray of carbendazim @ 0.1% at 50 and 70 days after sowing is recommended to manage the diseases.

Insects–Pests

Painted bug, sawfly and mustard aphid are the major insect-pests of the crop. Painted bug and mustard saw fly may be controlled by spraying of Endosulphan 35 EC. @ 0.035% (1 ml/lit. of water), spray of 0.025% solution of Metasystox 25 EC (1ml/lit. of water) is recommended for aphid control, at 15 days interval commencing from its first appearance in the crop.

Roguing

The Off-types should be removed from the seed plot before flowering. The objectionable weed Satyanashi (*Argemone maxicana*) must be rogued out from the seed field.

Harvesting and Threshing

When the siliquae begin turn to yellowish the crop should be harvested. After drying crop is threshed and care should be taken to avoid mechanical admixture.

Field Standards	F/S	C/S
Off-type (max.)	0.10	0.50
O.D.V. (max.)	0.10	0.50

Seed Standards

Pure seed (min.)	97%	97%
Inert matter (max.)	3%	3%
Other crop seeds (max.)	0.10	0.50
Total Weed Seed (max.)	0.10	0.50
Objectionable Weed Seeds (max.)	0.05	0.10
Germination (min.)	85%	85%
Moisture (max.)	8%	8%

LINSEED (*Linum usitatissimum* L.)

Linseed is normally a self-pollinated crop. Pollination takes place in mature buds before the flower open. Natural out crossing (0.5–0.9%) through bees or insects may occur. The flowers generally open in early morning. The flowering period of the crop is in the range of 25–40 days depending up on the temperature and moisture range during the flowering period.

Land Requirement

The selected seed field should be free from weeds and in the previous year the linseed crop should not have been taken.

Isolation

It is a highly self-pollinated crop. For seed production an isolation distance of 3 meters is required for foundation and certified seed, respectively.

Seed Rate

30–40 kg/ha.

Spacings

Line to line : 22 to 30 cm

Plant to plant : 5 to 6 cm

Fertilizers

Nitrogen 100 kg, phosphorus 60 kg and potash 40 kg/ha.

is recommended for linseed. Half quantity of nitrogen and entire quantity of phosphorus and potash should be applied as basal dose before sowing of the crop. The remaining quantity of nitrogen should be applied after first irrigation.

Plant Protection

Alternaria blight : Treat the seed with 2.5 g/kg. of Thirum before sowing. Sowing should be done in first week of November. In standing crop spraying of Mancozeb @ 2.5 kg/ha after 40–45 days of sowing and second spray may be done after 15 days interval.

White Rust : For the control of disease spray the crop with Mancozeb @ 2.5 kg/ha. For control of insects-pests spray the crop with Monocrotophos 36 EC @ 750 ml/ha or Endo-sulphon 35 EC @ 1.25 lit in 750–800 lit. of water per hectare.

Roguing

The Off-types and diseased plants should be rogued out from the seed field before flowering.

Harvesting and Threshing

When the capsules are ripe, the plant become brown, the crop is harvested. After drying the threshing of the crop is done and clean the seed.

Seed Standards	F/S	C/S
Pure seed (min.)	98%	98%
Inert matter (max.)	2%	2%
Other crop seeds (max.)	10/kg	20/kg
Objectionable weeds (max.)	—	—
Total Weed seeds (max.)	5/kg	10/kg
O.D.V. (max.)	10/kg	20/kg
Germination (min.)	80%	80%
Moisture (max.)	9%	9%

SOYBEAN (*Glycine max*)

Soybean has a typical papallionaceous flower with a tubular calyx of five unequal sepal lobes and a five-parted corolla consisting of posterior banner petal, two lateral wing petals and two anterior keel petals in contact with each other but not fused. The androecium consists of 10 stamens arranged in diadelphous manner. The elevated stamens form a ring around stigma. The pollination often occurs before the opening of flowers. The pollen is shed directly on the stigma. Due to presence of cleistogamy, there is a very high percentage of self-fertilization. The natural out crossing is generally less than 1%. The time from pollination to fertilization is 8–10 hours.

Land Requirement

This can be grown in all types of well drained soils. Such soils are not suitable which has <5 pH. Field selected should not have grown soybean in the previous season. The field should be free from volunteer plants.

Isolation

It is highly self-pollinated crop. Therefore, an isolation distance of 3 meters is sufficient for foundation and certified seed production.

Seed Rate

60–70 kg/ha is required. The seed should be treated with Rhizobium culture @ 375g/ha. The seed crop must be sown in rows.

Spacing

Line to line : 30 to 45 cm.

Plant to plant : 5 to 10 cm

Fertilizers

Apply full dose of nitrogen 25–30 kg, 50–60 kg phosphorus and 30–40 kg Potash per hectare and mix it well in the soil. If zinc deficiency is noticed, spray zinc sulphate and lime mixture on the crop.

Irrigation and Interculture

Irrigate the crop as and when needed depending up on soil and climate. The crop should not be allowed to suffer due to lack of water during flowering, seed filling and maturation stage. Weed should be controlled by one or two weedings. Weedicide such as Lasso @ 4 kg/ha. may be applied if needed.

Plant Protection

The major insect-pests are leaf eating caterpillar and pod borer. The important diseases are yellow mosaic and rust.

If the insect damage is observed, the crop should be sprayed with 1.75 ml Dimethoate 30 EC or 1 ml Methyl Parathion 50 EC or 2 ml Quinolphos 25 EC or 2 ml Endosulphon 35 EC mixed in 1 lit. of water. Alter natively the crop should be dusted with Parathion 2% @ 2.25 g/ha. or Malathion 5% @ 2.5 kg/ha.

For foliar diseases like Myrothecium, Cercospora leaf spot and Rhizoctonia aerial blight. Two sprays of Carbendazim 50 WP or Thiophanatemethyl 70WP @ 0.5 kg in 1000 lit. of water/ha at 35 and 50 days after sowing.

For control of rust a prophylatic spray of Hexaconazole or Propiconazole or Triadimefon @ 0.1% at onset of rust followed by subsequent sprays at 15 days interval.

Roguing

The seed plot should be monitored minutely through out the crop season specifically at flowering, pod filling and maturity stage. The Off-types are identified on the basis of varietal characteristics should be rogued out. The rogued plants should be removed from the seed plot.

Harvesting and Threshing

The soybean crop reaches harvestable maturity when the pods have lost their green colour and attain the mature pod colour characteristic of the variety and seed has become hard. The crop should be promptly harvested at this stage to avoid seed shattering and field deterioration. The soybean seed is highly prone to mechanical damage during harvesting if the seed moisture is below 13%. Therefore, desiccation should be avoided for the seed crop. After the few days drying when the seed moisture reaches 13–15% the crop should be threshed either by tractor treading or by a multi-crop thresher at 300–400 rpm. For direct combining, the moisture should be around 14%, the combine should be set carefully to avoid seed damage.

The processing should be carried out at seed moisture of 12–13%.

Field Standards	F/S	C/S
Plants of other varieties	0.10%	0.05%

Seed Standards	F/S	C/S
Pure seed (min.)	98%	98%
Inert matter (max.)	2%	2%
Other crop seeds (max.)	—	10/kg
Total weed seeds (max.)	5/kg	10/kg
Germination (min.)	70%	70%
Moisture (max.)	12%	12%
Off-types (max.)	0.10	0.20

GROUNDNUT (*Arachis hypogaea* L.)

The inflorescence of groundnut varies with different botanical varieties. The variety *Virginia* has simple inflorescence, expanding slightly in length during maturity, whereas in the Spanish they are compound and expand moderately. In *Valencia* the inflorescence is simple, but may elongate to form a conspicuous long branch that may occasionally terminate in leaves.

Groundnut flowers are typically papilionaceous and zygomorphic with a reduced pedicel. There are 10 monoadelphous stamens of which two are sterile represented only by filaments. The remaining eight are dimorphic, four long and four short.

Flower initiation takes place after 17–40 days after emergence depending on the genotype and environment. Generally the valencia are early in flowering and have short span of flowering. The spanish types also flower early, but the first flowering flush may be broader than in the *Valencia*. The *Virginia* types take more time than the other two types to start flowering.

Morphological differences among the Botanical Varieties of Cultivated Groundnut

Spanish (var. vulgaris)	Valencia (var. fastigiata)	Virginia (var. hypogea)
1. Branch (erect)	Branch (erect)	Semi spreading and spreading
2. Sequentially branched, 7–8 side branches	Sequentially branched; 4–5 tall branches	Profused branches, branches absent on main stem.
3. Stem-green tinge, pubescent	Purple	Green tinge, weak pubesence haulm as fodder.
4. Medium maturing	Early maturing	Late maturing
5. Leaves–smaller medium light green/dark green, elliptic, pointed tip.	Medium or large, round tip and base	Small leaves, dark green, inverted egg shaped, medium foliage
6. Uniform ripening of pod, pod setting compact.	Pods around main stem.	Bold pods
7. Pods-small 2 seeded	Medium-2–4 seeded	Large, 2 seeded
8. Testa colour-tan, red, white, or purple	Tan, fleshy red, white, yellowish, purple, variegated	Tan, fleshy red, white, yellowish, purple variegated.
9. Shell thin	Thick	Thin to thick.
10. Flowers on the axils of main stem	Flowers on main stem	No flowers on main axils, alternate 2 vegetative nodes with 2 reproductive.
11. Fresh seed dormancy usually absent	Fresh seed dormancy usually absent	Present upto 30–60 days.
12. Peg colour-green/ pigmented.	Green	Pigmented.

Flowers open normally at sunrise, but may be delayed by low temperatures. Anthers may dehisce 7–8 hours before flowers open. Stigma becomes receptive about 24 hour prior to anthesis and its receptivity persists for about 12 hours after anthesis. Fertilization occurs about 6 hour after pollination. The groundnut is strictly a self-pollinated crop. The natural out crossing is about 2 per cent.

Land Requirement

The crop does well on sandy loam, loam and well drained black soil. Select the field in which groundnut was not taken in two previous seasons. Area prone to bacterial wilt disease should be avoided.

Isolation

An isolation of 3 meter is sufficient for F/S and C/S seed production.

Seed Rate

100–125 kg Kernel/ha. bunch type and 80–100 kg/ha. in case of spreading varieties.

Crop can be sown during Kharif (end of June), Rabi (October–November) and in Summer (December–January).

Spacing

Row to row : 30 cm (bunch varieties); 45 cm (spreading varieties).

Plant to plant : 10 to 15 cm.

Fertilizers

The crop requires 20–25 kg nitrogen, 60–80 kg phosphorus and 25–35 kg potash per hectare. The entire dose of fertilizer is applied prior to sowing about 5-6 cm away from the seed row. The seed should be treated with Rhizobium culture @ 375 g./ha.

Irrigation and Interculture

The crop requires irrigation at an interval of 10–12 days depending upon the soil and climate. Adequate moisture at flowering, seed development and maturation is necessary.

One or two weedings are necessary followed by 2–3 hoeings.

Plant Protection

Major pests are aphid, leaf minor, red headed hairy caterpillar and grubs.

Major diseases are tikka, root rot, collor rot, rust and bud necrosis.

Leaf minor can be controlled by spraying 1 ml Monocrotophos or 0.5 ml Phosphomedon 85 WSC in one lit. of water. Spray solution needed is 600–800 lit./ha.

In case of aphid, spray Metasystox 25 EC (1 lit. dissolved in 1000 lit. of water/ha.). In case of leaf eating insect spray 2 ml Quinolphos 25 EC mix in 1 lit. of water. Use 600–800 lit of solution/ha.

In case of Tikka disease, spray 2 g. Chlorotheniol dissolved in 1 lit of water when the crop is 1, 1.5 and 2 months old. In case of rust, spray 2.25 g Mancozeb dissolved in 1 lit of water. Collor rot can be controlled by spraying Monocrotophos.

Roguing

The presence of Off-type plants *i.e.,* plants differing in their characteristics from those of the cultivar being grown is a potent source of genetic contamination. Roguing should be done at regular intervals (*i*) at about 30 days after emergence when most of the morphological characteristics of the cultivars have been expressed, (*ii*) at the time of flowering when flower characters have been expressed, and (*iii*) ten days before harvest when most of the foliar disease resistance characteristics become obvious.

Harvesting

The maturity of the crop is indicated by the yellowing of the leaves. The harvesting is done by pulling out plant by hand and allowed to be sun dried. The plants with pods strictly conforming to the pod characteristics of the variety in question should be bulked.

Seed Standards	F/S	C/S
Pure seed (min.)	96%	98%
Inert matter (max.)	4%	4%
Other crop seeds (max.)	—	—
Total weed seeds (max.)	—	—
Germination (min.)	70%	70%
Moisture (max.)	9%	9%
Off-types (max.)	0.10	0.20

Hybrid Sunflower

Sunflower is an important oilseed crop grown in the world as well as in India.

Land Requirement

It is cultivated in all types of soils including sandy soil. The field selected for hybrid seed production should not have been used to grow any sunflower crop in the preceeding season.

Seed and Sowing

(i)   Parental line seed production

A-line (Seed parent) – 4 kg/hectare

B-line (pollen parent) – 1.25 kg/ha.

R-line (Restorer) – 5 kg/ha.

(ii)   Hybrid seed production

Female (A-line) – 4 kg/ha.

Male (R-line) – 1.25 kg/ha.

Sowing

Dibble the seeds at recommended spacings, plant two seeds/hill and attend thinning to maintain one healthy seedling per hill, within 10–15 days after germination.

Row to row distance – 60 cm

Plant to plant distance – 30 cm

Staggered Sowing

Flowering behaviour of parental lines are ascertained in the hybrid seed production. Not more than 3–4 days difference in the flowering behaviour of parental lines (male and female) should be allowed to avoid stagger problems.

Isolation	F/S	C/S
Field of other varieties including commercial hybrid of the same variety	600 m	400 m
Field of the same hybrid not conforming to varietal purity requirements for certification and wild *Helianthus* spp.	600 m	400 m

Planting Methods

Two planting methods are being followed in sunflower to produce seeds.

Method I

For the breeder/foundation seed production of the female line, the planting ratio of A and B-line is 3 : 1. The sowing of A and B-lines should be taken by engaging labours separately and two seeds should be dibbled at each hill, at 3–5 cm in depth at a spacing of 30 cm with in the row and 60 cm between the rows. In respect to certified hybrid seed production A and B-lines are to be planted in 3:1 ratio and the management practices are similar as that of female seed production. In respect to B/S, F/ S seed production of the R-line, the entire plot should be sown with the respective restorer line and the same agrono-mic management has to be practiced.

Method II (Block Method)

The block system is suggested to avoid physical contamination by pollen shedders during production, harvesting, threshing and post-harvest handling operations. In this block method, A and B-lines in B/S and F/S stages as well as A and R-lines in case of C/S seed production are planted in a 75 : 25 ratio proportion using separate blocks. At the time of anthesis, the pollen is collected from B or R-lines and pollinated to A-line in respect of B/S, F/S and C/S production, respectively. This method ensures the production of high genetically pure hybrid seeds.

Fertilizers

80 kg N, 90 kg P_2O_5 and 60 kg K_2O per hectare is required. Apply ½ quantity of nitrogen and entire quantity of P_2O_5 and K_2O at the time of planting in furrows. The remaining half nitrogen may be applied as top dressing in two equal doses (I- 30–35 days and II- 50–55 days after sowing).

Roguing

In sunflower hybrid seed production plot, roguing is very important and has to be carried out in all stages of crop growth. The plants that do not conform to the respective parental lines have to be removed before anthesis. In F/S and C/S seed production, all the pollen shedders that could be easily identified by their dark anthers shedding pollen. In addition, plants with other morphological derivations such as in plant type, stem, leaf, height, hairiness, leaf size, margins colour and tip angles etc. are to be removed in A and B-lines before flowering. In the A and B-lines, the plants look alike for all the characteristics except sterility/fertility.

In B/S, F/S and C/S seed production plots of restorer lines, the Off-types have to be removed before anthesis. The roguing operation has to be carried out rigorously in all the parental lines from the vegetative stage and before commencement of flowering. The following traits are to be considered to identify the rogues :

❑ Plants that flower too early and too late should be removed.

❑ Plants showing morphological abnormalities like wrinkled and twisted leaves and deformed heads, should be rogued.

❑ Plants showing branching in non-branched lines should be rogued.

❑ Plants with an abnormal (pink or purple) colour should be removed.

❑ Plants with different colour seeds, seeds with varying streaks and other striped types (other than normal) should be removed at the time of harvest.

Field Inspection

A minimum of four inspections are to be carried out in hybrid seed production at four stages of crop growth :

1.	At 6-7 leaf stage	To determine isolation, out crossed volunteer plants, planting ratio, errors in planting, designated diseases and relevant factors.
2.	At flowering	Pollen shedders, Off-types are relevant factors.
3.	At maturity but prior to harvest	To verify designated diseases, true nature of plant and head characteristics of seed and other relevant factors.

Pollination

It is an important aspect to obtain better seed setting. It should be carried out through out the flowering period. This is done by gently passing the palm or palm covered with muslin cloth first on the 'B' plants and then by gently rubbing on the stigmas of the 'A'-line. In C/S seed production, the pollen of the 'R'-line has to be transferred to the A-line in similar way.

In the B/S and F/S of R-line production, hand pollination may be carried out by gently rubbing the plants with other plants. The hand pollination should be carried out in the morning hours (8 am – 12.00 noon) on all days through out the flowering period.

A minimum of four pollinations on 3rd, 5th, 7th and 9th days after anthesis should be carried out for better seed set. To avoid pollen theft by honey bees, it is better to provide natural smoke during the early morning hours (6am – 9am) or by spraying repellants on the crop.

Plant Protection

To control rust, leaf spot and downy mildew disease spray the crop with Dithane M-45 or Zineb @ 40g in 18 lit. of water when the crop is 40, 55 and 65 days old. About 600–800 litres of spray solution is needed per hectare. The plant affected by wilt and charcoal rot may uprooted and burnt.

Major insect-pests are Jassid, grasshoppers, leaf eating caterpillar, head borer and cut worm. Spraying during flowering may be avoided as it may affect pollinators.

Sucking insects can be controlled by spraying with Rogor 30 ml in 18 litres of water at 15–20 days interval as per incidence. Cut worm can be controlled by mixing 5% Heptachlor dust in soil @ 15 kg/hectare.

Special care should be taken to avoid bird damage particularly parrot. Scaring the bird by beating the drum and tins is the only control measure.

Field Standards	F/S	C/S
Off-types in seed parent at and after flowering	0.20%	0.50%
Off-types in pollinator at and after flowering	0.20%	0.50%
Pollen shedding heads in parent at flowering	0.50%	1.00%
Objectionable weed plants	—	—
Plants infected with downey mildew disease	—	—

Seed Standards	F/S	C/S
Pure seed (min.)	98%	98%
Inert matter (max.)	2%	2%
Huskless seed (max.)	2%	2%
Other crop seeds (max.)	—	—
Total weed seeds (max.)	5/kg	10/kg
Objectionable weed seeds (max.)	—	—
Germination (min.)	70%	70%
Moisture (max.)	9%	9%

SESAMUM (*Sesamum indicum* L.)

Land Requirement

The crop grows well on well-drained light loamy soil in Kharif and on medium heavy alluvial or black soil in Rabi season. The field should not have been under the same crop in the previous season.

Isolation

The crop is highly self-pollinated but it has recorded 4–5% cross pollination due to insect activities.

F/S   100 meters

C/S   50 meters

Seed Rate

2.5 to 5.5 kg/hectare

Spacing

Row to row spacing : 30 to 45 cm

Plant to plant spacing : 15 to 22 cm

Fertilizer Application

The fertilizer is applied @ of 35 kg nitrogen, 25 kg phosphorus and 25 kg potash at the time of sowing. Nitrogen may be applied in split doses. Half nitrogen is applied at sowing and the other half as top dressing at the time of first irrigation.

Plant Protection

Major insect pests are caterpillar, galfly and leaf-roller. They can be controlled by dusting 3% BHC dust @ 20 kg/hectare or spraying the crop with one ml Methyl parathion 50 EC or 0.05g Carbendazim dissolved in 1 lit. of water. The spray required is 500 litres/hectare.

Before sowing the seed must be treated with 3g mercurial compound disssolved in 10 lit. of water.

Roguing

Off-types and diseased plants affected by phyllody and leaf spot may be removed as and when they are found in the seed plot.

Harvesting and Threshing

The crop is harvested when the leaves, stem and capsules begin to yellow and the lower leaves start shedding. Before shedding the leaves, the crop is harvested by pulling or by cutting the plant. After complete drying of the plants the seeds are separated from the capsule by beating with the stick.

Field Standards	F/S	C/S
Off-types (max.)	0.10%	0.20%
Plants/head affected by designated diseases (max.)	0.50%	1.0%

Seed Standards	F/S	C/S
Pure seed (min.)	97%	97%
Inert matter (max.)	3%	3%
Other crop seeds (max.)	0.10	0.50
Total weed seeds (max.)	0.10	0.50
Germination (min.)	80%	80%
Moisture (max.)	9%	9%
O.D.V. (max.)	10/kg	20/kg

NIGER (*Guzotia abyssinica* L.)

Land Requirement

Selected field for niger was not grown in the previous season unless the crop grown was of the same variety and certified by Seed Certification Agency.

Isolation

Niger is a cross-pollinated crop.

F/S – 400 m

C/S – 200 m

Spacing

Row to row : 30 cm

Plant to plant : 10 cm

Fertilizers

The recommended fertilizer dose of 20 kg, N, 40 kg P_2O_5 and 20 kg K_2O per hectare may be applied just by the side of the seed row. Half nitrogen and full phosphorus and potash may be applied at the time of sowing. Remaining half nitrogen may be applied at the time of first irrigation.

Plant Protection

The crop suffers from leaf eating and bud eating caterpillar. Spray the crop with Methyl parathian 50 EC or Dimethiote 30 EC @ 500 lit. of water per hectare.

Roguing

All the Off-types and diseased plants should be rogued out as and when observed in seed crop.

Harvesting and Threshing

After maturity the crop should be harvested and threshed properly so that there could not be mechanical admixture in seed.

Field Standards	F/S	C/S
Off-types (max.)	0.05%	0.10%
Plants affected by seed borne diseases	—	—

Seed Standards	F/S	C/S
Pure seed (min.)	98%	98%
Inert matter (max.)	2%	2%
Other crop seeds (max.)	10/kg	10/kg
O.D.V. (max.)	10/kg	10/kg
Weed seeds (max.)	10/kg	10/kg
Germination (min.)	80%	80%
Moisture (max.)	9%	9%

HYBRID SAFFLOWER

Safflower has been under cultivation in India for Centuries for its orange red dye extracted from its brilliantly coloured florets. Nevertheless, it is preferred as a high value edible oil due to large concentration of polyunsaturated fatty acids. Use of safflower petals as a source of medicine and food colour is also gaining importance in recent times.

Selection of Field

Preferably select an area where safflower is not a regular commercial crop. In preceding two years the crop was not taken in that field where the seed production is to be organised.

Isolation Distance

The field selected for seed production should be isolated from other varieties of the safflower by 400 m and 200 m for foundation and certified seed production, respectively.

Planting Time

The best time for seed production is second fortnight of September to first fortnight of October.

Spacing

Row to row distance :   45 to 60 cm

Plant to plant :   20 to 30 cm

Principles for Seed Production

The effectiveness and efficiency of seed production programme largely depends on the production of parents, selection of effective technique and organization of hybridization.

There are three techniques to produce hybrid seed in safflower :

1. Emasculation and hand pollination

2. Mass emasculation

3. Male sterility system.

The first two techniques are mainly used in breeding programmes and the third technique involving both genetic and cytoplasmic genetic male sterility systems is largely used to produce hybrid seed.

In safflower, the GMS system is being effectively used for hybrid seed production in large scale. The first safflower hybrid, DSH–129 and MKH-11 have been developed using GMS.

Male Sterility System in Safflower

In safflower both CGMS and GMS are being exploited for hybrid development in the crop. However, the most prevalent system for development of hybrids in safflower is Genetic male sterility (GMS) system.

1. **Single recessive GMS :** GMS sources governed by single recessive gene being exploited for hybrid development are :

    (*i*)   UC–148 and UC–149 GMS lines developed by Heaton and Knowles (1980).

    (*ii*) CMS male sterile lines developed by Ramchandran and Sujatha (1991).

(*iii*)   MSN and MSV male sterile lines developed by Singh (1996).

(*iv*)   DMS male sterile lines associated with dwarfness developed by Singh (1997).

Each of these male sterility sources segregates in the ratio of 1 MS : 1 MF plants. The identification of sterile and fertile plants is possible at 30–40 days after sowing. At the age of 30–40 days, MS plants (remain dwarfs of 5-10 cm height. However, MF plants attain a normal height of 20–25 cm. This height difference facilitates roguing of MF plants at this stage itself leaving behind 100% pure plant stand of dwarf ms plants.

2.   **Dominant Genetic Male Fertility :** In safflower the only source of male sterility governed by a single dominant gene is reported by Joshi *et al.* (1983). Owing to dominant nature of male sterility causing gene, the MS lines as well as hybrids based on this source segregate 1 MS : 1 MF plants. In this type too identification of MF plants is possible only at flowering stage.

3.   **Cytoplasmic Genetic Male Sterility :** The exploitation of CGMS for hybrid development in safflower has been carried out in the U.S. (Hill, 1989). However, lack of suitable restorer for the sterile cytoplasm seems to be a major hurdle in commercial scale exploitation of CGMS for hybrid development in safflower, as all the hybrids based on this system exhibited a lack of fertility restoration in them (Suing *et. al.* 2000). Therefore, it appears that the CGMS system which is available in U.S. is facing some basic problems which need to be tackled before it can be exploited commercially in safflower.

Selection of Inbreds

1.   Only pure lines are used as a male parent for hybrid development programme.

2.   Pure lines are crossed in diallel manner to know their gca effects.

3. Lines with good gca effects for yield and yield components are crossed with genetic male sterile lines for identification of promising hybrids.

Selection of Genetic Male Sterile Lines

1. Line should be highly uniform for different traits like plant height, days to 50% flowering, capitulum size and shape.
2. There should be better opening of the male sterile capitula to allow exertion of stigma out of florets for effective pollination.
3. Male sterile line should have high yielding ability.
4. Better responsiveness to fertilizer and irrigation.

F_1 Hybrid Seed Production

1. Similar to the seed production of CMS line, the foremost requirement of isolation distance of at least 1000 meters for other safflower plots.
2. Row proportion of 6 GMS : 2 male parent is followed.
3. Spacing of male and female parent is 45 cm between rows and 20 cm between plants.
4. GMS line (female parent) segregates in 1 : 1 ratio of MS and MF. Rogued out all the MF plants 5–6 days prior to flower initiation.
5. Any Off-type appearing in female or male parent rows is removed immediately.

Diagrammatic Representation of Hybrid Seed Production

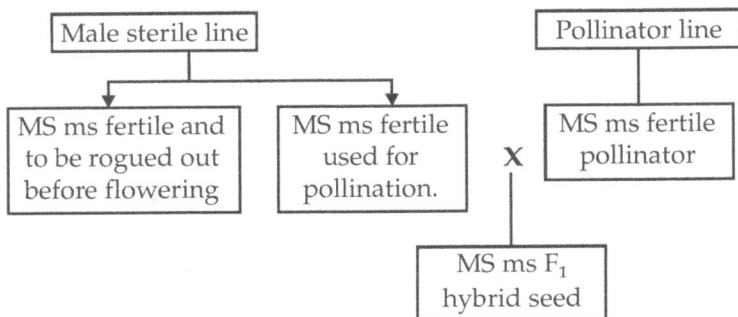

Male sterile line		Pollinator line
MS ms fertile and to be rogued out before flowering	MS ms fertile used for pollination.	MS ms fertile pollinator

X

MS ms F_1 hybrid seed

Field standards	Minimum permitted %	
	F/S	C/S
Off-types		
(a) male parent	0.05	0.10
(b) female parent	0.05	0.10

Seed Standards	F/S	C/S
Physical purity (min.)	98	98
Inert matter (max.)	2	2
Other crop seeds (max.)	—	—
Total weed seeds (max.)	5/kg	10/kg
Objectionable weed seed (max.)	—	—
Germination (min.)	80	80
Moisture (max.)	9	9
O.D.V. (max.)	—	—

REFERENCES

Heaton, T.C. and Knowles, P.F. (1980). Registration of UC-148 and UC-149 male sterile safflower germ plasm *Crop Sci.*, 20 : 554.

Hill, A.B. (1989), Hybrid safflower breeding. In : Proc: 2nd Intl. Safflower Conference, Hyderabad, India, 9-14 January, 1989. Indian Society of Oil seeds Research, DOR, Hyderabad, 169-170.

Joshi, B.M.; Nerkar, Y.S., and Jambhale, N.D. (1983). Induced male sterility in safflower, *Journal Maharashtra Agric. Univ.*, 8 : 194-196.

Ramchandran, M. and Sujatha, M. (1991). Development of genetic male sterile lines in Safflower. *Indian J. Genet.*, 51 : 268-269.

Singh, Vrijendra (1997). Identification of genetic linkage between male sterility and dwarfness in safflower. *Indian J. Genet.*, 57 : 327-332.

Singh, V.; Desh Pande, M.P.; Galande, M.K.; Deshmukh, S.R. and Nimbkar, N. (2000). Current status of research and development in Safflower hybrid in India. *Natl. Seminar on Oilseeds Res. and Dev.* : Needs in the Millenium, ISOR, DOR, Hyderabad, February 2–4, 2000, pp. 62.

HYBRID CASTOR

Castor is normally monoecious with pistillate flowers on the upper portion of the raceme and staminate flowers on the lower part. The proportion of female flowers in different order racemes is an important attributes which has strong bearing on ultimate yield. For higher yields, an ideal sex phenotype of a variety or hybrid should have female flowers right from the base of raceme with staminate flowers confined to bottom one or two whorls sufficient to provide required pollen.

Since castor is highly cross pollinated crop, out crossing between mostly female and mostly male parents lead to rapid regression to mostly male sex phenotypes in subsequent generations. One way to minimize such high sex instabilities is to expose the basic seed stock of varieties and male parents to male promoting environment.

Isolation

The extent of cross pollination is mainly depends on the direction and velocity of wind which is the primary source of pollen dispersal in castor.

	F/S	C/S
Male parents	1000 meters	600 meters
Female parent	1500 meters	1000 meters

Planting Time

Time of planting and specific season has profound influence on sex expression. While summer and Kharif season provide ideal male promoting environment for undertaking seed production of varieties, male and female parent of hybrid. Rabi winter is the most ideal season for taking up hybrid/certified seed production as it is most conducive for production of female flowers. In case of varieties and male parents such an exposure to male promoting environment *i.e.,* Kharif and summer encourages good expression of less productive plants bearing mostly male spikes which could be easily eliminated through timely roguing. Similarly, the female parents when raised in male promoting environments produce environmentally sensitive staminate flowers which are very essential for self reproduction of the female parents.

Spacing

Line to line : 90 cm

Plant to plant : 30 cm

Seed Rate

Female : 3 kg/ha.

Male : 2kg/ha.

Fertilizers

Fertilizer is applied at the time of sowing @ 80 kg N, 60 kg P_2O_5 and 25 kg K_2O/hectare. Half dose of nitrogen and entire dose of P_2O_5 and K_2O are applied as basal and remaining half quantity of nitrogen should be given in two split doses *i.e.* (45 to 60 days after sowing and ¼ after first picking).

Sowing

The male lines are sown first in border in all the directions. Then the female lines are sown in 3 rows alternating with one row of male line.

Roguing

Off-type plants and diseased plants should be removed as soon as possible when they are spotted in the field before flowering.

Harvesting and Threshing

The capsules start ripening in November to December and continue doing so till March-April due to uneven maturity. The bunches are removed when 3–10 capsule start turning yellow. Threshing is done by beating with a mallet and the seeds are separated by winnowing.

Maintenance of Parental Lines (Male Parent)

Source Seed :   Selfed bulk seed of selected progenies

Season :	Summer of Kharif.
Isolation :	1500 m
Agronomic Practices :	Follow all recommended practices.
Roguing :	Visit regularly the B/S production plots for detection and elimination of plants which do not conform to the designated morphological characters of genotypes.

❑ Exert selection pressure for environmentally sensitive sex expression.

❑ Allow the male flowers only to the bottom 2–3 whorls.

❑ At the end of 2nd round of roguing, ensure that at least 50% of the population has similar node number of general mean and the rest are less or more than the mean.

❑ Before first picking, identify four representative samples of atleast 100 random plants each and label them. Record all the morphological characters, yield contributing characters and oil content. Estimate the population means, standard deviation and PCV for yield and yield components.

❑ Harvest, thresh and process the seed in the pickings at 25–30 days interval starting from 100–120 days.

❑ Keep the picking – wise produce separately and store under well ventilated ambient conditions.

Female Parents

Source of Seed :	N/S
Season :	Summer (Second fortnight of January)
Isolation :	2000 meters.

Rogue out all monoecists atleast 2–3 days earlier to flowering in the primary raceme. Verify individual female plants for various morphological characters, particularly the number of nodes upto primary raceme. Most of the female flowers on primary raceme fail to set fruits consequent to non-availability of pollen.

A most of interspersed late male flowers however, sprout on primary as well as subsequent order racemes in about 35–50% female populations which provides sufficient pollen for the later developed female flowers on the same raceme as well as later sequential order racemes. Observe all plants regularly for any reversion to monoecism upto 4th order raceme and rogue out. However, the pistillate plants reverting the monoecism in 5th sequential order onwards can be allowed in the population and supplement pollen source. Collect the seed from all female plants and keep the picking wise seed lots separately after proper drying.

Field standards	Max. Permitted (%)	
	F/S	C/S
Off-types including plants found to flower over the main stem	0.50%	1.00%
Male variants (three secondary rachis from base and above possessing only flowers are considered as mostly male	1.00	2.00
Female variants (in certain cases all sex variants of dominant females, females with interspersed staminate flowers mostly female and mostly male raceme spectra involved	1.00	2.00
Monoecious plants and the racemes reverted to monoecism on female plants before anthesis	—	2.00

Seed Standards	F/S	C/S
Pure seed (min.)	98%	98%
Inert matter (max.)	2.00	2.00
Other crop seeds (max.)	—	—
Weed seeds (max.)	—	—
O.D.V. (max.)	5/kg	10/kg
Germination (min.)	70%	70%
Moisture (max.)	8%	8%
Genetic purity	95%	95%

7

SEED PRODUCTION OF FIBRE CROPS

HYBRID COTTON (*Gossipium hirsutum* L.)

Land Requirement

The land should be free from volunteer plants of cotton. There should be good drainage because it does not tolerate water logging.

Isolation

Cotton is self-pollinated crop but it has recorded natural cross pollination to the extent of 10–50%. An isolation distance of 50 meters for foundation seed and 30 meters for certified seed production is required.

Seed, Season and Sowing

Treated parental seed should be obtained from approved source. The best season for planting is July 15th to August 15th.

The seed production can be taken in two ways :

1. **Planting Ratio Method**

   The area under parental lines should be approximately 3:1 to 5:1.

2. **Planting by Block System**

   Male and female parents are planted in two separate blocks. The main seed plot consists of female parent only. The male parent should be planted in the adjacent plot separately.

Since the flowering period in cotton is spread over a long time, the sowing of the area reserved for the male parent is done in 3–4 instalments as shown below in a ratio 60 : 20 (female to male).

1. First 4 rows of male along with female rows.

2. Next 6 rows after 8 days.

3. Next 6 rows after 16 days.

4. Final 4 rows after 24 days.

The purpose of staggered planting is to make available pollen for complete crossing programme.

Seed Rate

2–3.75 kg/ha. female

1–2.50 kg/ha. male.

Spacing

In case of female parent

Row to Row : 120 to 150 cm

Plant to Plant : 90 to 100 cm

In case of male parent

Row to Row : 90 cm

Plant to Plant : 60 cm

Fertilizers

A fertilizer dose of 125–150 kg nitrogen, 50–75 kg phosphorus and 50–75 kg potash/hectare is usually recommended. Apply full dose of P_2O_5, K_2O and ½ dose of nitrogen as basal dressing. The remaining 50 per cent nitrogen may be split into 3–4 equal doses applied as top dress at 30–35, 40–45 and 75–80 days after sowing as per the seed production experimental results.

Foliar Application of Fertilizers and Nutrients

(i) Diammonium phosphate application – Four to five sprays of 15% DAP (15 g DAP in 1 lit of water) may be sprayed during boll development for better development of crossed bolls.

(ii) NAA @ 10 ppm (4.5 ml in 18 lit. of water) may be given at square initiation (40–65 DAS), flower initiation (60–70 DAS) and peak flowering (90–110 DAS) for reducing the physiological shedding.

Weeding

Keep the seed production plot weed free throughout the growth stages.

Irrigation

Irrigate the crop only when it is necessary with in 40–45 DAS.

Plant Protection

(i) Seed treatment with Imidachloprid 70 WS @ 10 g or Thiomethoxam 70 WS @ 5g/kg seed.

(ii) For management of white flies spraying of Trizophos 1.5 ml/lit or Phosalone 35 EC 1.5 ml/lit can be followed.

(iii) Dicofol 20 EC 2.5/lit or Wettable sulphur 5.5 g/lit can be used for management of mite incidence.

Bollworm Management

For management of bollworms spraying of Carboryl 50 WP (4.0 g/lit.), Endosulfan 35 EC (2.8 ml/lit.), Quinolphos 25 EC (2.5 ml/lit.), Chlorpyriphos 20 EC (2 ml/lit.), Profenohos 50 EC (2.5 ml/lit.) and Novaluran (1ml/lit.) Avoid repeatation of same chemical any time.

The activity of bollworms can be monitored using pheromone traps. For monitoring Heliothis incidence install the traps around 50 DAS and for PBW around 80 DAS. Change the lures of each trap at 15–20 days.

Avoid growing of jowar, maize, groundnut, sunflower, chickpea around seed production plots to reduce bollworm damage.

Important Steps in Seed Production

1. Off-plants of male and females should be removed before commencement of crossing programme.

2. Emasculate and pollinate maximum buds.

3. Pollination may be done between 8 am to 1 pm.

4. Choose optimum size of the bud.

5. Cover the male buds with paper bag on the previous evening. Collect the buds early morning before bud opening.

6. Emasculated buds of females may be covered preferably with butter paper bag.

7. Close the crossing programme after 10 weeks after initial flowering.

8. Crossed bolls collected in baskets.

9. Insect damaged bolls should be sorted out.

10. Male plant should be removed on the dates specified by Certification Agency.

11. Every day unemasculated flowers and flower buds in females should be removed.

12. Clean the seed perfectly.

Seed Standards	F/S	C/S
Pure seed (min.)	98%	98%
Inert matter (max.)	2%	2%
Other crop seeds (max.)	5/kg	0.10
Weed seeds (max.)	5/kg	0.10
Germination (min.)	65%	65%
Moisture (max.)	10%	10%
Off-type (max.)	0.10%	0.20%

Ginning

It is the last operation and has a big role to play in cotton seed quality. The cotton should be ginned in a ginning machine which is approved by the Seed Certification Agency. Here utmost care has to be taken to keep the ginning machine clean to avoid admixture and also should be worked properly to avoid mechanical damage. The seed should be dried to safe moisture level.

JUTE (*Corchorus olitorius* and *C. capsularis*)

Both the species of jute are predominantly self-pollinated crops. Average cross-pollination of *capsularis* jute is about 1.38% and that of *olitorius* jute is about 10.5%. The amount of out crossing ranges from 1.04–16.04 per cent in *capsularis* and 3.30–17.70 per cent in *olitorius* depending upon the pressure of pollinating agents and weather conditions.

Usually in both the species of jute 2–3 flowers open at a time on any inflorescence and flowering is completed with in a period of 4–6 weeks in *capsularis* and prolonged to about 10 weeks in *Olitorius*. Flowers open between 8.30 am to 10.30 am and close at 3.30 pm.

Land Requirement

Jute needs light, sandy loam, preferably in the areas, where facilities for protective irrigation exit. The land should be fertile and well drained and of neutral pH. The field selected should not have been under jute cultivation in the previous season.

Isolation

In case of foundation and certified seed production the minimum isolation should be 50 and 30 meters, respectively.

Spacing

Row to row : 30 cm

Plant to plant : 10 cm

Seed Rate

8–10 kg/ha. (*Capsularis*)

4–6 kg/ha. (*Olitorius*)

Fertility

In *capsularis* varieties 80–100 kg nitrogen, 20–30 kg P_2O_5 and 20–30 kg K_2O/ha. is recommended. Half nitrogen and full dose of P_2O_5 and K_2O is applied before sowing. Remaining half quantity of nitrogen should be given in equal dose after 3–4 weeks and 6 weeks days after sowing. In case of *Olitorius* variety 60–80 kg nitrogen is required. The quantity of P_2O_5 and K_2O and method of application is same as in *Capsularis*.

Roguing

Roguing is carried vigorously at three stages of plant growth :

1. 30–40 days after sowing

2. At flowering stage

3. At capsule formation stage

At all these stages, the Off-types, diseased and pest infested plants should be removed from the seed plot.

Plant Protection

Indigo caterpillar can be controlled by dusting BHC 10% @ of 12–15 kg/ha. For other insects spraying with Folidol E 605 @ 1CC/5 lit. of water is required to control.

The major diseases are anthracnose, soft rot, die back, leaf mosaic, mildew. For anthracnose control, spray the crop with copper oxychloride. For soft rot, the crop should be sprayed with Bordeaux or copper oxide.

In case of leaf mosaic, the infected plants should be uprooted and destroyed.

Harvesting and Threshing

When the capsules become brown, the crop is ready for harvesting. Harvesting should be done in morning to avoid shuttering losses.

Field Standards	F/S	C/S
Off-types (max.)	0.50%	1%
Plants affected by seed borne diseases (max.)	1.0	2.0

Seed Standards		
Pure seed (min.)	97%	97%
Inert matter (max.)	3%	3%
Other crop seeds (max.)	10/kg	20/kg
Weed seeds (max.)	10/kg	20/kg
Germination (min.)	80%	80%
Moisture (max.)	9%	9%

MESTA (*Hibiscus cannabinus* L.)

Land Requirement

The crop comes up well on light black soil and sandy alluvial loams. It is not suited to water logged conditions.

The field selected should not be taken mesta crop in the previous year.

Isolation

100 meters – foundation seed.

50 meters – certified seed.

Spacing

Line to line : 20 to 30 cm

Plant to plant : 10 to 15 cm

Fertilizers

70–80 kg   Nitrogen/ha.

120–140 kg P_2O_5/ha.

98 kg K_2O/ha.

Half quantity of nitrogen and full quantity of phosphorus and potash should be applied as basal. The remaining half nitrogen should be supplied as top dress in split doses.

Roguing

Off-typed plants should be rogued from the seed field whenever they are observed. The plants affected by stem rot and root rot should be removed as soon as they are observed.

Harvesting and Threshing

The harvesting should begin when the crop is matured otherwise there is shattering of the seed. The crop is cut by sickle and allowed to dry in the field for 3–4 days.

Threshing is done by hand or thresher. After cleaning the seeds should be dried to safe moisture level.

Seed Standards	F/S	C/S
Pure seed (min.)	98%	98%
Inert matter (max.)	2%	2%
Other crop seeds (max.)	20/kg	0.1%
Weed seeds (max.)	20/kg	0.10
Germination (min.)	80%	80%
Moisture (max.)	9%	9%
Off-types (max.)	0.50%	1.0%
Plant affected by designated diseases (max.)	0.10%	0.20%

8

SEED PRODUCTION OF FORAGE CROPS

BERSEEM (*Trifolium alexandrinum*)

Berseem is predominantly an out cross crop. The honey bee plays significant role in pollination and as a result, affects the seed production.

The stamens are diadelphous (9 + 1) and stigma globular. Anthesis takes place early in the morning but the pollen grains are enclosed with in the corolla, as the flowers open at later stage. The mechanism is simple valvular in which the stamens are forced out of the keel under pressure by an insect, but return to the original position when the pressure is released. Presence of bee coloney *Apis dorsata* and *A. meliferra* increases seed production significantly.

Land Requirement

Berseem grows well on well-drained rich loamy soil. It can tolerate moderate alkalinity but not acidity in the soil.

The land selected should not have been under the same crop during previous two years. The land should be free from volunteer plants.

Isolation

Foundation seed : 400 meters

Certified seed : 100 meters

Sowing

Where berseem is sown in a field for the first time, it is necessary to inoculate seed with special bacterial culture. For inoculation, mix one kg of culture in 1 lit. of 10% sugar solution. Seed is then mixed well with the solution and then dried under shade.

The field is irrigated and in the deep standing of 5–7 cm, the seed are broadcast @ 25–30 kg/ha.

Irrigation

Initially, 2–3 irrigations are given every 7–10 days depending upon the nature of soil for good germination and early establishment of seedlings. The subsequent irrigations should be given at an interval of 15 days during winter.

Plant Protection

Spray Endosulfon 0.5 per cent to control lucerne Caterpillar, semilooper and lucerne weevil. The spray mixture required is 800–900 lit/ha. For the control of thrips spray 0.05% Diazinon or 0.02% Metasystox solution.

Cuttings

After February–March cutting, the crop should be left for seed production. Timely last cutting will ensure insect activity, pollination and also avoid lodging of the crop.

Roguing

There should be complete removal of Off-types, other crop plants and chicory plants at all stages of crop growth.

Harvesting and Threshing

When majority of the pod turn brown, the crop should be harvested and after drying in field for 3–4 days, threshed and clean.

Field Standards	F/S	C/S
Off-types (max.)	0.10%	0.20%
Objectionable weed plants	0.05	—

Seed Standards	F/S	C/S
Pure seed (min.)	98%	98%
Inert matter (max.)	2%	2%
Other crop seeds (max.)	0.10	0.50
Weed seeds (max.)	0.10	0.50
Objectionable weed seed (max.)	5/kg	10/kg
Germination (min.)	80%	80%
Moisture (max.)	10%	10%

LUCERNE (*Medicago sativa* L.)

Land Requirement

The land should be free from volunteer plants.

Isolation

400 meters : F/S

100 meters : C/S

Sowing

Sowing can be done as berseem in respect of inoculation.

Seed Rate

20–25 kg/ha.

Fertilizers

A basal dose at the time of sowing with 25–30 kg nitrogen, 50 kg phosphorus and 25 kg potash/ha is sufficient.

Cutting

Cutting the crop for fodder should be stopped by the end of February.

Roguing

All the Off-types, other crop plants and dodder plants should be removed before harvesting. Thorough roguing should be done at all stages of the crop growth.

Harvesting and Threshing

When the pods turn brown harvest the crop and after proper drying threshing is performed.

Seed Standards	F/S	C/S
Pure seed (min.)	98%	98%
Inert matter (max.)	2%	2%
Other crop seeds (max.)	0.10	0.50
Weed seeds (max.)	0.10	0.50
Objectionable weed seed (max.)	5/kg	10/kg
Germination (min.)	80%	80%
Moisture (max.)	10%	10%
Off-type (max.)	0.10	0.20

9

SEED PRODUCTION OF ROOT CROPS

Vegetable crops play vital role in vegetarian diet by way of providing minerals, vitamins and fibre. Among vegetable crops, the root crops are considered good source of vitamin C (radish, turnip and beet root), vitamin A (carrot) and minerals like Iron (Radish, turnip and beet root), calcium (beet root and turnip) and phosphorus (beet root). The major root crops are carrot, radish, turnip and beet root which are commonly grown in India.

RADISH (*Raphanus sativus* L.)

Radish is a very popular root crop throughout India. It is grown for its fleshy edible roots which are eaten raw or as salad or cooked. It belongs to family cruciferae. There are two distinct genetical groups in radish.

(*i*) Asiatic varieties (*ii*) European varieties.

Land Requirement

Well drained sandy loam soils are best suited for the crop. The soil should be thoroughly prepared so that there are no clods to interfere with the development of the root. The plot should be free from volunteer plants.

Isolation

It is a cross-pollinated crop and requires more isolation from other varieties to avoid any out crossing with regard to purity of seeds.

	Mother root production		Seed production	
	F/S	C/S	F/S	C/S
Field of other varieties	5 m	5 m	1600 m	1000 m
Field of same variety not Conforming varietal purity	5 m	5 m	1600 m	1000 m

Seed field should not be located too close to clover crop due to its high bee activity.

Seed, Season and Sowing

For seed production, a less humid climate is suitable. Temperature of 33°C or higher cause the stigma to become dry and pollen grains fail to germinate. The sowing time in radish is depends upon time required in root formation/root development and methods of seed production. Among the Asiatic varieties *i.e.,* Pusa chetki and Pusa Rashmi should be sown in September and middle of October, respectively. These varieties develop good root within 40–45 days after sowing. The Japanese white which takes 55–60 days in root development should be sown in first week of August and Pusa Himani can be sown in middle of August. In case of seed to seed method the sowing can be delayed than root to seed method.

Seed Rate

8–10 kg/ha is sufficient for the production of roots.

Selection of Roots for Seed Production

Fully developed roots are lifted along with tops and arranged in rows to select the true to type roots. The roots are selected based upon the size, shape and pithiness. The normal roots after twisting the top are put in the bucket of water. Those roots with a degree of pithiness are discarded and the solid roots are retained. While in case of round rooted variety only tap root is removed before planting. The selected roots are immediately transplanted in the field in both types *i.e.,* Asiatic and biennial.

Spacing

The 4 inch size stecklings are transplanted in prepared field with a spacing of 45–60 cm between the rows. The distance between the plants should be maintained at 8–10 cm. The field should be irrigated immediately after transplanting.

In hills, the transplanting is done in last week of October, while in plains transplanting should be done in first fortnight of November.

Fertilizers

During the first season, 20–25 tonnes of FYM/ha is mixed with soil which is followed by the application of 40 kg nitrogen, 50 kg phosphorus and 50 kg potash/hectare before sowing. Remaining 40 kg nitrogen is applied as top dressing when the roots start elongating.

During IInd season, similar quantity of FYM is applied at the time of land preparation for transplanting purpose. At the time of transplanting, apply 50 kg phosphorus and top dress with 25 kg nitrogen per hectare in the pre-bolting stage and another dose of 25 kg nitrogen just before the start of flowering.

Methods of Seed Production

There are two methods of seed production

In situ method (seed to seed)

Ex. situ method (roots to seed)

In seed to seed method, the crop is allowed to over winter in the same field and produce seed in the following spring in their original position. However, in root to seed method, during autumn the roots when fully developed are lifted and selection of true to type is made. Deformed, under developed, split and Off-types roots are rejected. The roots are also examined for pithiness. The tap root is pruned and tops are clipped with out damaging the crown shoots. Selected roots are transplanted immediately in Asiatic as well as in temperate types.

Irrigation and Interculture

Irrigate the crop at an interval of 8–10 days depending upon the requirement of the crop and weather.

Weeding once and once earthing during early stages of the crop growth will ensure proper development of the roots.

Plant Protection

The major pests and diseases are aphid, saw fly and downey mildew. Spray the crop with 20 ml Malathion or 40 g Carbaryl in 10 lit. of water two weeks after sowing. About 315 lit. of mixture may be required per hectare. Repeat the spray 4 weeks after sowing.

Spray 20 g Mancozeb or 30g Difolton in 19 lit. of water for the control of downey mildew. Repeat the spray after two weeks. Use about 450 lit. of spray mixture/hectare.

Harvesting and Threshing

The crop is harvested when fully mature and allowed to dry on the threshing floor. Threshing can be done by beating the sticks. The seed is cleaned and dried to safe moisture levels.

Inspections

First inspection at 20–30 days after sowing of the crop for isolation distance and Off-types.

Second inspection at the time of root development/root lifting. To check the root shape (long/globe), colour (white, red, yellow, black, bicolour), pithiness, number of leaves at maturity, relative height and leaf shape.

Third inspection at flowering to check the isolation, Off-types and designated diseases like black rot and black leg.

Roguing

Remove Off-types and diseased plants at all inspection stages.

Field Standards	F/S	C/S
Plants of other species	0.10%	0.20%
Diseased plants	0.10%	0.50%

Seed Standards	F/S	C/S
Pure seed (min.)	98%	98%
Inert matter (max.)	2%	2%
Other crop seeds (max.)	5/kg	10/kg
Weed seeds (max.)	10/kg	20/kg
Germination (min.)	70%	70%
Moisture (max.)	6%	6%

TURNIP (*Brassica rapa* L.)

The turnip belongs to family Cruciferae. The roots contain vitamin B and C in appreciable quantities. It has hermaphrodite flowers and is highly cross-pollinated mainly due to self-incompatibility nature. It is a herbaceous annual for the root production, where as it is biennial for the seed production. The crop has two distinct groups (*i*) Asiatic types (*ii*) European types.

Climate and Soil Requirements

Asiatic varieties are sown earlier, requiring warmer conditions than the European types. Turnip required sandy loam to clay loam soil and it should be rich in organic content. It can also grow in a soil having pH between 5.5 to 6.8.

Sowing Time

The sowing of Asiatic type is done from July to September, where as the temperate types are sown last week of August to first week of September.

Seed Rate

3–4 kg seed is sufficient to produce the roots for transplanting 6–10 hectares of land.

Spacing

Sowing on ridges 45 cm apart is preferred to flat sowing because ridge facilitate better root development. Thinning should be done in order to maintained 4–5 cm spacing between plants in root production.

Weeding and Earthing Up

One hand weeding at early stage of growth and earthing up to promote the good root development.

Methods of Seed Production

Both seed to seed and root to seed are employed for seed production. The seed to seed method gives better yield than root to seed method. However, root to seed method is used for production of breeder/nucleus seed.

Preparation of Steckling Root

During November fully developed roots are lifted and selected for true to type character. The under developed, deformed and Off-type roots are rejected. After pruning the tap root and clipping of top leaving brown intact the root are planted in freshly developed field at distance of 45–60 cm. The stecklings are transplanted immediately in Asiatic and temperate varieties.

Topping

Topping is done to encourage the development of secondary inflorescence on the main roots. This reduces the over all height of the crop and possibility of lodging at flowering and maturity. The top 10 cm of the terminal shoot is removed when flowering shoots are between 30–40 cm height.

Isolation

	Mother root production		Seed Production	
	F/S	C/S	F/S	C/S
Field of other varieties	5m	5m	1600m	1000m
Field of same variety not conforming the varietal purity	5m	5m	1600m	1000m
Other spp. of genus *Brassica*	5m	5m	1600m	1000m

Roguing

Off-types and diseased plants should be rogued out as and when they appeared in seed plot.

First Inspection at 20–30 days after sowing in order to check the isolation and Off-type.

Second Inspection at root development stage or at root lifting. At this stage root shape (Flat, globe or long globe) leaf shape, colour and relative size, pithiness/spongyness etc. examined.

Third Inspection at flowering verify the isolation distance, Off-types and designated diseases like black rot and black leg.

Flowering, Pollination and Fruiting

The Asiatic varieties start flowering in last week of December under *in-situ* while 10–15 days later in *ex-situ* and continued till February. In hills among temp. varieties flowering begins from March and crop is in full bloom in April. The inflorescence is a typical raceme and chief pollinating agent is honey bees. To ensure good seed set, pollination of all the flowers on inflorescence can be achieved by keeping the bee hive adjacent to seed field.

Harvesting and Threshing

Turnip has a tendency of shattering therefore care is required in cutting. It is therefore, suggested to cut the whole crop when 60–70 (%) of pods turns yellow brown in colour. The crop is cut with sickle and left in windrows until the seeds are mature. The turning upside down is practiced to cure the material. Drying of crop is essential for easy threshing. The threshing is done by beating with a stick or by thresher. The seed should be dried under sun thoroughly prior to cleaning of seed.

Seed Standards	F/S	C/S
Pure seed (min.)	98%	98%
Inert matter (max.)	2%	2%
Other crop seeds (max.)	5/kg	10/kg
Weed seeds (max.)	10/kg	20/kg
Germination (min.)	70%	70%
Moisture (max.)	6%	6%

BEET (*Beta vulgaris* L.)

The garden beet is a member of the family Chemopodiaceae. It is a wind cross-pollinated crop having protandrous flowers. The garden beet is different from Sugar beet. The true seeds are small, kidney-shaped and brown. It produces seed in the temperate climate.

Climatic and Soil Requirements

It is primarily a cool-season crop but grows well in warm weather and hence can be grown during winter all over the plains. It thrives best on a fairly deep friable loam, moist but well drained soil. It is very sensitive to acidic soil but thrives very well in alkaline soils with pH as high as 9–10.

Isolation Distance

	Mother root production		Seed Production	
	F/S	C/S	F/S	C/S
Field of other varieties	5m	5m	1600m	1000m
Field of same variety not conforming the varietal purity	5m	5m	1600m	1000m
Field of Garden beet and Sugar beet	5m	5m	1600m	1000m

Sowing Time

The sowing is done from mid-July to end of July in hills. The late varieties are sown from the end of June to mid-July.

Seed Rate

8–10 kg seed is sufficient for one hectare of land which are sufficient to plant 2–3 hectare area under seed crop.

Thinning

Seedlings are singled as soon as possible following emergence to allow to root development.

Spacing

The planting distance should be 60 cm between rows and 45–60 cm within the rows.

Steckling Size

The optimum size of steckling for transplant should be 2.50–2.75 cm in diameter and weighing 40–45 g.

Method of Seed Production

Singh *et. al.* (1960) reported that root to seed is usual method of seed production in Kullu valley. In this method, during November–December the well developed roots are dug out. After selection the tap roots are trimmed without damage to crown and planted in a well prepared field.

Pollination, Flowering and Seed Setting

Beet root is predominantly wind pollinated. Bolting in beet root start in first fortnight of April in Kullu valley and crop is in full bloom from mid-May to mid-June. The inflorescence is a large panicle and seed maturity begins from the base of panicle.

Roguing

Off-type plants should be rogued out from the seed plot.

Harvesting and Threshing

The crop matures in July in Kullu valley. However, when the summer is hot and dry therefore harvesting may start as early as the last week of June, but usually it is done from the first week of July to end of July. Generally, when 70–80% of seed balls on plant gets hardened and those at the base of the inflorescence turn brown, the crop is harvested. Delaying in harvesting may lead to shattering of seed during harvesting. The seed crop is then stacked for curing and then dried under sun.

Threshing is done manually by beating with stick or by tractor treading.

Plant Protection

The important insect-pests of beet root are leaf minor, web worm and Semi loopers. For control of leaf minor spray of any Systemic insecticide. For the control of web worm spraying the crop with Rogor @ 1ml/lit. of water and for semi loopers Endosulphan 20 EC spraying.

For control of leaf spot disease spraying of crop with Blitox 50 (0.3%) and for the control of downey mildew Dithane Z–78 (0.3%).

Seed Standards	F/S	C/S
Pure seed (min.)	96%	96%
Inert matter (max.)	4%	4%
Other crop seeds (max.)	5/kg	10/kg
Weed seeds (max.)	5/kg	10/kg
Germination (min.)	60%	60%
Moisture (max.)	8%	8%

CARROT (*Daucus carota* L.)

Land Requirement

The field selected for carrot seed production should not have any volunteer plant. It can be grown on well drained fertile loam and sandy loam soil.

Isolation

	Mother Root Production		Seed Production	
	F/S	C/S	F/S	C/S
Field of other varieties	5m	5m	1000m	800m
Field of same varieties not conforming the varietal purity	5m	5m	1000m	800m

Seed, Season and Sowing

For the seed production the crop is considered biennial because in the first season the plants form roots. In the second season roots are planted and they produce flowering shoot, flowers and form seed. The seed is produced by two methods (1) root to seed (2) and seed to seed method.

Generally the first method is followed because in the second method the crop suffers widely due to root rot infection.

Root to Seed Method

During autumn the roots when fully developed are lifted and selection of true to type is made. Under developed, deformed, split and Off-types are rejected. The roots are also examined for core size and colour. The tap root is pruned and tops are clipped without damaging the crown shoots. Selected roots are transplanted immediately in the field.

Spacing

Row to row : 75 cm
Plant to plant : 6–7 cm   } For seed sowing on ridges.

Fertilizers

At the time of land preparation, 15–20 tonnes FYM/ha is mixed well with soil. Fertilizer dose of 40 kg nitrogen, 50–60 kg phosphorus and 125 kg potash/ha. The crop is top dressed twice @ 20 kg nitrogen/ha. At the time of transplanting apply 50–60 kg nitrogen, 50–60 kg phosphorus and 50–60 kg potash/ha and mix well with soil. Apply 50–60 kg of nitrogen/ha after hoeing and earthing up.

Irrigation and Interculture

Depending upon the soil and climatic condition, the crop may be irrigated once in 10 days. Two to three hand weedings are required followed by hoeing in the early stages of crop growth. It should be followed by earthing.

Transplanting

Selected roots are transplanted at a distance of 25–30 cm and field is immediately irrigated.

Flowering and Seed Setting

The Asiatic types tend to be annual and flowers during March in plains. However, bolting in temperate cultivars occur in first week of April and flowering starts by end of May in hills. The individual carrot flowers are normally protandrous and much pollination occurs between plants in seed crop. However, because of the extended flowering period resulting from several successive umble/plant and the succession of flower in individual umble, the possibility of selfing is always remains there. The anthesis in single umbel is completed in 7–9 days. The peripheral umbellete flowers first followed by inner umbelletes. Bohart and Nya (1960) observed the occurrence of pollinating insects on flowering carrot and noted that honey bees were efficient pollinators. They also observed that several insect gevea are Hymenoptera, Diptera and Cleoptera were extremely important pollinators of carrot, in an absence of bees. The pollination activities were greater at 10 am and were closely related to temperature. Howthorn *et al.*, (1950) found that when there was a adequate supply of pollinating insects both seed yield and seed quality were high where natural insect pollinators population were low then there was an advantage in increasing the number of honey bee colonies by placing group of hives adjacent to this carrot field.

Seed Setting

Seed setting is influenced by the position of umble, order of umbel bees activities and the temperature at flowering. Seed setting is 100% and 82.70% in primary and secondary umbels, respectively as observed by Singh *et.al.,* (1960) in cv. Imperater. However, the seed setting is very poor in 3rd and 4th orders umbels due to non-availability of pollen being desiccation of pollen caused by high temperature.

Plant Protection

The major pests are leaf hopper, weevil and aphid and among the diseases the major ones are leaf spot, blight and powdery mildew.

Spray the crop with 20 ml Malathion or 40 g Carboryl with 24 g wettable sulphur and 30 g copper oxychloride in 10 lit. of water, 4 weeks after sowing. About 360 lit. of spray mixture is required for 1 hectare area. Repeat the spray, 7th and 10th week after sowing.

Roguing

Small, diseased, cracked or injured roots may be discarded. Early bolters and Off-types should be removed from time to time.

Harvesting and Threshing

The seed mature in the middle of May and end of June in plains and hills, respectively. The harvesting of umbels should be done where the secondary umbels are fully ripe and third umbels have started to turn brown. For high quality seed primary and secondary umbels should be harvested and rest should be avoided. The umbel is cut manually without taking the stem part. The harvesting is done in the early morning to take an advantage of dew to reduce the loss from dropping.

The carrot seed has spines or beards and these must be removed by debearder before further cleaning operation.

Field Standards	F/S	C/S
Off-type plants (max.)	0.10%	0.20%

Seed Standards		
Pure seed (min.)	95%	95%
Inert matter (max.)	5%	5%
Other crop seeds (max.)	5/kg	10/kg
Weed seeds (max.)	5/kg	10/kg
O.D.V. (max.)	5/kg	10/kg
Germination (min.)	60%	60%
Moisture (max.)	8%	8%

References

Bohart, G.E. and W.P. Nye (1960). Insect pollination in Carrot Utah. Bulletin 419, Agri. Experiment Station, Utah State Uni.

Howthon, L.R. (1950). Studies on soil moisture and spacing for seed crop of Carrot and Onion. USDA, Circular No. 892.

Singh, H.B.; M.R. Thakur and P.M. Bagchandani (1960). Indian J. Hort., 17 : 38–47.

POTATO (*Solanum tuberosum* L.)

Land Requirement

Well drained, weed free and free from tuber borne diseases and neutral to slightly alkaline in reaction soil is suitable for seed production of potato. It is better to take up potato after 2–3 years of rotation.

Isolation

An isolation of 5 meters is required for foundation and certified seed production.

Seed, Season and Sowing

The sowing season is between last week of September to the second week of October. Seed rate mainly depends upon tuber size. The tuber should have bold buds ready to sprout or under sprouting for quick emergence. Traditionally small size

(30 g) tuber are used. But the optimum seed size for seed production purpose is 30–40 g for plains and 50–60 g for hills. Dip the seed tuber in a solution of Mencozeb @ 4g in one lit. of water for 5 minutes and dry in shade before sowing to prevent decay of the seed tubers.

Spacing

Furrows are made at 50 cm. Plant the tubers in furrows at 20 cm. More than 90 per cent area is planted manually. Opening shallow furrows and applied fertilizers to the furrows and mix with soil. Place the tubers and make the ridges with spade or bullock drawn ridger.

Fertilizers

Apply 20–25 tonnes of FYM 2–3 weeks before and mix it well with the soil. A fertilizer dose of 100 kg nitrogen, 100 kg phosphorus and 100 kg potash/hectare is required. At the time of planting half quantity of nitrogen and entire quantity of phosphorus and potash should be applied. Remaining half quantity of nitrogen is given at the time of earthing. Fertilizers should be applied 5 cm away from the tubers.

Irrigation and Interculture

Apply irrigation after 20–25 days after sowing. Frequent, light irrigations at low moisture tensions is must for faster root development. During formation of stolens and tuber initiation, water stress limits the number of tubers and during enlargement phase, it depresses the size of tubers (Hukkeri *et al.*, 1970). Tuber initiation was reported to be critical stage and even single irrigation at this stage tend to increase yield (Singh *et al.*, 1975).

Hand weeding should be taken up once or twice to keep the field free of weeds. Earthing up should be taken when the plants attain a height of 15 cm.

Haulms Cutting

Haulms should be removed from 5th to 15th January when *M. persicae* population reaches 20 aphids/100 compound leaves.

Haulms should be cut by sickle to the ground or by spraying of Paraquat @ 3.0 lit./ha. Vines should be put on the ridges to ward off the adverse effect of high temperature and direct sun light on the exposed tubers. Exposed tuber should be covered and regrowth must be checked by spraying Paraquat @ 2.5 lit./ha one week after haulm cutting.

Plant Protection Measures

At the time of planting Phorate 10G @ 10 kg/ha should be applied to control jassids, leaf hoppers and white flies. At earthing up apply 7–5 kg/ha. Phorate 10G. Two sprays with Dimethoate 30 EC or Methyl demeton 25 EC @ 1 lit./ha. in the form of Metasystox or Rogor may be given at an interval of 12–15 days depending up on the duration of crop and appearance of aphids.

For early blight and late blight one prophylatic spray with Dithane M–45 @ 2 kg/ha and it should be repeated 10–14 days interval depending upon the weather conditions.

For cut worms, ridges should be drenched with Chlorpyrifos 20 EC @ 2.5 lit./ha and for leaf eating caterpillars or defoliators, Endosulfan 30 EC. @ 1.5 lit/ha should be sprayed.

Roguing

Aphid, *Myzus persicae* is the most important virus vector for potato. Among potato viruses, leaf roll and virus Y are very serious as due to their infection tuber yield reduced by 25–80% (Nagaich, 1974).

For quality crop of seed potato, roguing has to be done at :

1. Before earthing up
2. 40–50 days after planting
3. Just before haulm cutting/removal.

These three inspections are must for roguing mosaics, leaf roll and Off-type plants. Plants showing symptoms of veinal necrosis, crinckle, mosaic and marginal yellowing should be removed.

Harvesting

The crop is ready for harvest 10–15 days after dehaulming when skin is hard enough to withstand handling operation. During digging, care should be taken to avoid injuries to the tubers. After lifting, the tubers should be removed to well ventilated place without exposing it for longer time in the field. They are kept in piles for about a week to get rid off the excess moisture there by further hardening of the skin of the tuber takes place. Digging can be done by tractor or bullock drawn digger or by spade manually.

Grading

Grading is essential and graded tubers should be washed in clean water and then dip in 3% solution of boric acid for 30 minutes to control scabs, rhizoctonia and other surface borne diseases. Treated tubers should be dried in shade and sent to cold store in first week of March and latest by end of March.

Field Standards	F/S	C/S
Plants of other varieties (max.)	0.05%	0.1%

TPS (True Potato Seed) Production

TPS is a low cost alternative of producing high quality planting material and the cost of cultivation can be reduced by 15–35% at farm level through TPS utilization.

TPS is a viable and successful propagule, for potato production (Walker and Crissman, 1966). Exploitation of TPS technology potentially overcome some of the problems associated with seed tubers and has proven advantageous over the conventional method of potato cultivation using tuber seed.

1. As much as 18% of the total produce of edible potatoes used as seed can be saved for domestic consumption.

2. Cost of planting material can be reduced sharply as a handful of TPS (approx. 100 g) could replace upto 2–3 tonnes of perishable, bulky and costly seed tubers required to plant one hectare of land area.

3. Spread of tuber–borne diseases can be minimized which, except one or two viruses and one viroid (Jones, 1982) are not transmitted through TPS.

4. TPS can be stored easily for many years at room temperature with a little loss in viability (Simmonds, 1968).

Heterotic effect for yield in TPS can be obtained when *Andigena* clones are bred with *Tuberosum* clones. Maximum heterosis for tuber yield has been reported by *Tuberosum* × *Andigena* crosses both in F_1 seedling and F_1C_1 tuber generations (Dayal, 1981). Female parental lines selected from *tuberosum* carry resistance for late blight from *S. demissum* and the male parental lines carry resistance from *S. andigena*. High pollen production and fertility in the male lines and moderate pollen fertility together with the self-incompatibility in female lines are desired for higher berry/seed set and production of hybrid TPS of desired quality and quantity with low inputs.

In the production of TPS, the effect of altitude (climate), photoperiod and manipulation of female parent, are observed on berry set, size of berry, number of seed/berry and quality components of TPS. There is a differential response of the female parent depending on the genotype. In general, high altitude, cooler climate and 14–16 hours photoperiod favour better blooming, higher berry setting, larger size of berry with higher seed number coupled with higher proportion of A-type seed.

Suitable parents can be selected for cultivation which can yield high quality TPS of an hybrid or OPV. The yield can be increased by the application of suitable fertilizer doses and also by reducing the competition within and between flowering branches.

Potato has a multiovular gynoecium containing 1000–1200 ovules, out of which only 30–40% of the ovules generally get fertilized and produce seed.

The stigmatic surface has been observed to show changes in its surface structure in relation to receptivity. The pollen load directly determines the quality and quantity of TPS produced.

Hence, there is a need to pollinate three times within the receptivity period of 18–36 hours of the stigma to produce high yield of TPS.

The cost of production can be decreased by resorting to pollination without emasculation.

The crop raised through TPS is almost disease free.

TPS are also having dormancy. The seed soaked for 4 days either in 1 per cent KNO_3 or 2 per cent KH_2PO_4 and dried back for 2 days eliminated the dormancy.

Seed Standards	F/S	C/S
Off-type (max.)	0.05%	0.1%
Mild mosaic	1.0%	3.0%
Mosaic and leaf roll	0.5%	1.0%
Root knot nematode	—	—
Wart	—	—
Brown rot	—	—
Scab	0.10%	0.50%

Tubers should not be less than 2.5 cm in average diameter about 30 g in weight.

References

Dayal, T.R. (1981). Study of heterosis in Potato and the use of some induced tetraploids for its exploitation. Ph.D. Thesis, Agra University, Agra, India. pp. 191.

Simmonds, N.W. (1968). Prolonged storage of potato seed. *Eur. Potato J., 11* : 150–156.

Walker, T. and C. Crissman (1996). Case of the economic impact of CIP. related technology CIP, Lima, Peru. p. 157.

10

SEED PRODUCTION OF FRUITS & VEGETABLES

OKRA (*Abelmoschus esculentus* L.)

Lady's finger or Okra is cultivated throughout India for its immature fruits which are generally cooked as vegetable. It belongs to family Malvaceae with bisexual flowers.

Land Requirements

It grows best in comparatively lighter soils ranging from sandy loam to loam and have proper drainage. The field selected should not have been cropped with Okra in the previous season.

Isolation

Though it is a self-pollinated crop, natural crossing does take place to the extent of 5–20 per cent. The seed plot should be isolated at least by 400 meters in case of foundation seed and 200 meters in case of certified seed production.

Seed, Season and Sowing

The crop can be sown in Kharif, Rabi and Summer. Usual dates of sowing are June-July, September-October and January-February. But summer crop is good for seed production.

The seed rate per hectare required is 8–10 kg. After the land is ready for sowing, prepare ridges and furrows 45–60 cm apart. The seeds are soaked in water for 15 hours before sowing for getting better germination. The distance between plant to plant should be kept at 30 cm.

Fertilizers

The fertilizer requirement per hectare is 120–125 kg nitrogen, 60–75 kg phosphorus and 50–60 kg potash. Half quantity of nitrogen and entire dose of phosphorus and potash should be used as basal dressing. Remaining half dose of nitrogen is applied one month after sowing as a top dressing.

Irrigation

Usually the crop should be irrigated once in a week. In the rainy season, the crop needs no irrigation. When there is a prolonged drought then irrigation is to be given.

Plant Protection

Spray the crop with 20 ml Oxydemeton methyl or 5 ml Phosphomedon in 20 lit. of water two weeks after sowing. Use about 350 lit. of spray mixture per hectare.

After 5 weeks of sowing, spray the crop with 20 ml Malathion in 10 lit. of water. About 450 lit. of water is required per hectare. Repeat this spray after 7 weeks of sowing.

For the control of powdery mildew, spray 30 g wettable sulphur in 10 lit. of water. Repeat the spray after 15 days if it is not controlled. Use about 450–550 litres of spray per hectare.

In case of cercospora leaf spot, spray the crop with 10g Benomyl or 10g carbendazim in 10 lit of water. Repeat the spray after 15 days if disease still existed.

For the control of yellow vein mosaic the affected plants should be removed immediately and crop can be sprayed with 18 ml Dimethoate in 10 lit. of water.

Roguing

The Off-types and wild Okra plants should be removed from the seed field before flowering. The plants affected by yellow vein mosaic should be removed and destroyed.

Harvesting

When the capsules have dried, should be harvested manually and threshed. Dried the seeds upto safe moisture level.

Field Standards	F/S	C/S
Off-types (max.)	0.10%	0.20%
Plants affected by yellow mosaic (max.)	—	—

Seed Standards	F/S	C/S
Pure seed (min.)	98%	98%
Inert matter (max.)	2%	2%
Other crop seeds (max.)	—	5/kg
Weed seeds (max.)	—	—
Objectionable weed seed (max.)	—	—
Germination (min.)	65%	65%
Moisture (max.)	10%	10%
O.D.V. (max.)	10/kg	20/kg

BELL PEPPER (*Capsicum annum* L.)

Bell Pepper commonly known as sweet pepper or capsium is one of the most popular and highly remunerative vegetable crop grown throughout the world. Capsicum is a self-pollinated crop but the extent of natural out crossing has also been reported upto 66.40 per cent. (Vanangamudi *et al.*, 2003), which is due to high insect activities.

Land Requirement

Well drained loam soils rich in organic matter, sandy or sandy loam soils are preferred where growing season is short. Acid soils and water logging conditions are not suitable.

Planting Time

In the regions of mild winter and moderate monsoons capsicum can be grown all the year round. Although crop can be grown at any time of the year, April and June is the best time.

In South India best season is from October to February. The main seed crop sowing in May, June and transplanted in July.

Seed Rate

1–1.5 kg/ha. Four–six weeks old seedlings are ideal for transplanting.

Fertilizers

115 kg nitrogen, 200 kg phosphorus and 200 kg potash is required for one hectare area. Application of 20 kg Borax and 20 kg calcium carbonate per hectare proved best in recording highest yield.

Irrigation

An adequate supply of irrigation water must be available because capsicum are prone to flower drop. The droping off under developed fruits occur during times of water stress, more so with the long fruited types.

Roguing

Roguing has to be attended at vegetative, early flowering stage and at maturity.

Before Flowering

Desirable characters – Growth habit, vigour and foliage typical of the cultivars, leaf characters, observe any specific seed borne diseases.

Early Flowering and First Fruit Immature

General plant habit and characters checked at stage 1. Observe for specific seed borne disease.

Mature Fruit

General plant habit and characters checked at stage 2, fruit colour when ripe, fruit size, shape and length.

Isolation

Capsicum is self-pollinated crop due to allogamous nature, out crossing is upto the extent of 58–76 per cent.

Foundation seed – 200 meters.

Certified seed – 100 meters.

Use of Growth Regulators

NAA (Naphthalene Acetic Acid) is promising growth regulator for increasing yield. NAA at 50 ppm was found to increase yield.

Harvesting

The best time of harvest is at 50–60 days after planting when the fruits are green bright to deep red. Red ripe fruits are picked, cut and macerated mechanically to separate the seeds.

Seed Standards	F/S	C/S
Pure seed (min.)	98%	98%
Inert matter (max.)	2%	2%
Other crop seeds (max.)	5/kg	10/kg
Total weed seeds (max.)	5/kg	10/kg
Germination (min.)	60%	60%
Moisture (max.)	8%	8%
Off-type plants (max.)	0.10%	0.20%
Diseased plants (max.)	0.10%	0.50%

References

Vanangamudi, K.A. Bharathi; N. Natarajan and V. Mallika (2003). New technologies in seed industry. *Agro India,* May–July, 2003.

SEED PRODUCTION OF HYBRID TOMATO

Tomato (*Lycopersicon esculentus*) is one of most popular and extensively consumed vegetable. It is self-pollinated crop belongs to family solanaceae. The phenomenon of hybrid vigour in tomato was first observed as early as in the beginning of 20th century.

Earliest reports on heterosis was made by Hedrick and Booth (1908), Willington (1922), were the first to advise the use of F_1 hybrid for boosting tomato production.

Hybrid seed production technology involves three important steps :

1. **Production of inbred lines :** Selection and selfing of the lines is generally adopted from germplasm having wider genetic diversity.

2. **Testing the combining ability :** The combining ability of different genotypes was tested. The lines with high GCA effects were selected and make crosses. On the basis of sca effects the best combinations were selected based on the performance.

3. **Production of F_1 hybrid seed :** Tomato is self-pollinated crop. The flower favours complete self pollination without the need of insects.

Sowing

Sow the male parent upto 8–10 days earlier than the female parent to ensure an adequate supply of pollen for fertilization of female plants. The ratio of male to female plants approximately 1:5.

Seed Requirement

Female (100 g) and male (25g) seed is required for planting of one hectare seed plot. Ideal sowing season for seed production is October-November.

Spacings

Line to line 50 cm (male), and 40 cm (female).

Isolation

An isolation distance of 200 meters for foundation (parental lines multiplication) and 100 meters for certified class (hybrid) is required.

Steps in Hybrid Seed Production

The production of hybrid tomato seed involves three steps *viz.* (*i*) Emasculation (*ii*) Pollen collection (*iii*) Pollination

1. Emasculation begins about 55–65 days after sowing depends upon duration of female line. The flowers on the female lines must be emasculated during their late bud stage for pollination. Buds should be plumpy fully developed. These buds are emasculated 12–14 hours before opening.

2. Before crossing both lines should be checked for true to type and undesired suspected plants should be removed.

3. Pollen is collected from male parent in petry-dish covered with cloth bag and shake well for release of pollen grains.

4. Collected pollens should be applied with a fine brush to each emasculated female flower. Pollination should be done from 9 a.m. to 12 noon. Each pollinated flower is marked with coloured tag.

Anthesis and Fertilization

Pollination one day after emasculation resulted maximum fruit setting. Emasculation and pollination simultaneously carriedout, resulting poor seed set.

Staking and Pruning

The female parent should be staked irrespective of their growth habit. Among male lines, only indeterminate types need to be staked.

Roguing

Thorough roguing should be done at following stages :

I. *Seedling stage :* presence/absence of anthocyanin in stem, uniformity, leaf morphology.

II. *Before flowering* : growth habit, leaf characters, general habit, diseases.

III. *Early flower and fruit stage* : general habit, immature fruit green back fruits, ridges or without ridges.

IV. *Fruiting stage* : fruit colour, size, stage.

Use of Male Sterile Lines

The cost of Hybrid seed can be reduced by the use of male sterile lines. Four types of male sterility have been reported in tomato :

1. Pollen sterile type (Pollens are not functional)

2. Functional male sterility (pollens are functional)

3. Stamen less flower

4. Positional sterility (protruding stigma).

Harvesting

Tomato fruit ripen about 50–60 days after pollination, but may take longer if temperature are cool. Keep the fruits on vine until they are fully mature, preferably to the pink or red ripe stage. Be sure to check for the clipped sepal before harvesting fruit. Collect fruits in nonmetallic containers such as nylon net bags, plastic buckets or crates.

Seed Extraction

Seed extraction may be done manually or mechanically. Crush the fruits with trampling with feet. Put the bags of crushed fruits into big plastic containers and ferment to separate the gel mass embedding the seed. To hasten the fermentation process, put weight over the bags or keep the fruits submerged in the liquid fruit mass. The time of fermentation depends upon the ambient room temperature. If temperature is above 28°C. One day of fermentation may be sufficient. Fill up the container with water and stir the seeds to allow the pieces of flesh and skin sticking on the seeds to float. Incline the container and gently remove the floating refuse, making sure that the seeds remain at the bottom. Repeat the washing several times by adding fresh water to the container every time until all the flesh and gel are completely removed, leaving clean seeds at the bottom.

Mechanical seed extraction is used by large scale operations. Instead of fermentation, treat the seed gel mass with HCl (hydrochloric acid) @ 7 ml/kg of seed-gel mass. Stir the seed-gel mass while the acid is being added. Continue stirring for 40 minutes until the gel is visibly softened or dissolved. When the seed is separated from the gel, pour the acid-treated seeds into a clean fine-mesh bag. Wash the bag with tap water thoroughly so that no acid is left on the surface.

Seed Drying

Place the wash seeds in fine mesh bags. Excess water can be removed by hanging the seeds in the shade for a day. After the excess water is removed, uniformly spread the partially dried seeds in a flat plastic container and place in a seed dryer. Drying continues for 3–4 days, maintaining a temperature of 28–30°C. Well dried seeds should be graded and treated with halogen mixture @ 3g/kg. seed.

Field Standards		
Off-types in seed parent	0.05%	
Off-types in Pollinator parent	0.05%	
Pollen shedders in seed parent	0.10%	
Plants affected by seed borne diseases	0.50%	

Seed Standards	F/S	C/S
Pure seed (min.)	98%	98%
Inert matter (max.)	2%	2%
Other crop seeds (max.)	5/kg	10/kg
Weed seeds (max.)	—	—
Germination (min.)	70%	70%
Moisture (max.)	8%	8%
Genetic purity (mini.)	90%	90%

SEED PRODUCTION OF HYBRID BRINJAL

Brinjal (*Solanum melongena* L; Solanaceae; 2n=24), ranks second in terms of area and production, next to potato under Indian conditions.

Land Requirement

Land to be used for seed production of hybrid brinjal shall be free of volunteer plants, well drained and highly fertile besides free from soil borne pests and diseases.

Nursery Activities

Male and female entries are raised separately on nursery beds. Male entries have to be sown 8–10 days well in advance as compared to female. Transplanting is done when the seedlings are 25–30 days old.

Main Field

Male has to be transplanted 8 days well in advance as compared to female. Male and females have to be transplanted in separate blocks.

Spacing

Line to line : 75 cm

Plant to plant : 60 cm

Isolation

	F/S	C/S
Field of other varieties including commercial hybrid of the same variety	200 m	200 m
Field of same hybrid not conforming varietal purity	200 m	200 m
Between blocks of parental lines in case female (seed) parent and pollinator are planted in separate blocks	—	5m

Roguing

Before flowering : Plant habit, foliage typical of the cultivar, angle of leaf, leaf pigmentation and spinyness. *Early flowering and first fruit development.* General plant habit, vigour and characters as defined for stage, flower colour and any other marker.

Fruiting

Shape, size, colour and patterning when ripe, calyx colour and spinyness.

Emasculation

Emasculation is done only in female line. Emasculation is carried out three days earlier to pollination of that flower, when in the floral buds the anther colour just starts turning yellow. The calyx are removed at the time of emasculation. After emasculation butter paper bagging is done to prevent cross pollination. Emasculation begins after 2 pm and continues upto 4 pm.

Pollination

Pollens are collected from male parent, when the anthers colour turn dark yellow and the flower is about to open. Pollens are collected on the previous day of pollination in the afternoon (after 2 o'clock). If there is sunshine then flowers are exposed to sunshine and the pollens collected from them will be more in quantity and have better and more viability.

Pollination will be carried out in morning between 8 am – 12 noon and tagged the crossed flower. Leaving 10–15 fruits/plant, all the flowers and fruits are completely removed before development.

Harvesting

Ripened fruits are harvested by hand plucking. Fruits are cut with a instrument and squeezed to a mass of pulp and seeds (the pericarp being thrown out). The pulp is removed immediately by repeated washing with fresh water as it floats and the seeds are thoroughly dried under shade by spreading them thinly on thin nylon net above the muslin cloth.

Field Standards	F/S	C/S
Off-types in female parent (max.)	0.010%	0.050%
Off-types in male parent (max.)	–	0.050%
Plants affected by seed borne diseases (max.)	0.10%	0.50%

Seed Standards	F/S	C/S
Pure seed (min.)	98%	98%
Inert matter (max.)	2%	2%
Other crop seeds (max.)	—	—
Weed seeds (max.)	—	—
Germination (min.)	70%	70%
Moisture (max.)	8%	8%
Genetic purity (mini.)	99%	90%

HYBRID CHILLI

Chilli (*Capsicum annum* L.) is an important vegetable and spice crop of the world. Being rich in vitamins, especially vitamin C, chilli fruits are put to various edible use through out the world.

Flower is solitary, extra axillary sometimes occurs in pairs. Flower is abractiate, actinomorphic, pedicellate, bisexual and hypogynous. Calyx is capanulate. Androecium consists of 5 stamens, epipetalous, alternating with corolla lobes, anthers, apparently connate, often opening by pores. Gynoecium consists of 2 carpels, syncorpous, superior ovary. Style is slender, terminal, linear, Sigma is sub capitate and faintly bifid.

Flowers are open at 5 am – 6 am. The stigma is receptive from a day earlier to anthesis and continues for 2 days after anthesis.

Anther dehiscence normally starts at 9 am and may continue upto 11 am. Pollen grains are fertile a day before anthesis with maximum fertility on the day of anthesis.

Maintenance of parental lines : Chilli being often cross-pollinated crop and pollinated by insects.

1. Hand selfing – Artificial hand pollination.
2. Caging the plants with fine meshed cotton/nylon cloth.
3. Bagging a side branch before opening of flowers with perforated butter paper bag.
4. Selfing (hand pollinating) the unopened buds a day prior the anthesis.
5. Growing in isolation.

Isolation Distance

A minimum of 500 meters isolation is required for maintaining the genetic purity.

Roguing

All the Off-type and diseased plants should be rogued out from seed plot.

Hybrid Seed Production

The hybrid seed can be produced by the use of :

1. Male Sterility
   (a) GMS
   (b) CMS
   (c) CGMS
2. Self-incompatibility.
3. Chemical hybridizing agent
4. Manual emasculation and pollination

However, in Chilli commercial hybrid seed production is done predominantly by manual emasculation and pollination and also by the use of CGMS system.

For hybrid seed production, flower buds which are going to open next day only are to be selected. Emasculation may be done either in the previous evening with the help of a forcep, the petals are carefully parted and anthers are removed and bagged. In the afternoon of the following day, fresh flowers should be plucked from the male parent which have been previously bagged and pollen dusted on to the stigma of the emasculated flower buds.

Harvesting

The fruit should be picked when red ripe. After drying the seed should be separated and cleaned.

CUCUMBER (*Cucumis sativus* L.)

The cucumber is an important summer vegetable in all parts of India and is used as salad, as pickle and also as cooked vegetable. It is a warm season crop and it does not withstand even light frost. The cucumber varieties are generally classified into four groups :

1. European-American.

2. West Asiatic

3. Chinese

4. Indo-Japan

Land Requirement

The cucumber is grown successfully on many kinds of soil from sandy to heavy clay. The cucumber plant grows well at soil reaction between pH 5.5 and 6.7. The soil should be well drained. There are no particular requirement about the previous crop since individual seed discrimination is not possible as they are held in a fruit.

Isolation

It is highly cross-pollinated crop. An isolation distance of 1000 meters for foundation seed and 500 meters for certified seed is required.

Seed Rate

2.5 to 5 kg/hectare.

Spacing

Row to row : 1.5 to 2.0 meters.

Plant to plant : 60 to 90 cm.

Fertilizers

The total requirement of the crop is 55 kg nitrogen, 45 kg phosphorus and 85 kg potash per hectare.

Sex Expression and Sex Ratio

The question of sex expression and sex ratio is of great interest in cucumber, which have monoecious plants. These bear male and female flowers separately on the same plant. Generally there are more male flowers and female flowers bear fruits.

Gibberellic acid at lower concentrations of 10–25 ppm increases the number of female flowers. Two sprays, one at 2 leaf stage and again at 4–leaf stage with 25–100 ppm of Maleic hydrazide, 100 ppm of alpha-naphthalene acetic acid, 200 ppm of Ethrel, 3 ppm of boron or 3 ppm of molybednum can suppress the number of male flowers and increase the number of female flowers, fruit set and ultimate yield.

Roguing

All the Off-type and diseased plants should be rogued out from the seed plot as and when they appear.

Insect-pests and Diseases

Cucumber are attacked by a number of diseases. The most important diseases are bacterial wilt, anthracnose, downey and powdery mildew, angular leaf spot and mosaic. The most important insects are red pumpkin beetle, aphids, cut worms and fruit fly. Adopt the suitable control measures for the control of insect-pests and diseases.

Harvesting

When the cucumber attains dark brown colour, they should be plucked. The seeds from the fruits are extracted by acid or alkaline method. 8.5 ml HCl per 25% technical grade ammonia per 1000 kg of freshly threshed material is thoroughly stirred into pulp. After about 30 minutes, water is added with stirring. The pulp and other impurities will float and mature seeds sink to the bottom. The seeds are then dried to safe moisture content.

Field Standards	F/S	C/S
Off-type plants (max.)	0.10%	0.20%

Seed Standards		
Pure seed (min.)	98%	98%
Inert matter (max.)	2%	2%
Other crop seeds (max.)	5/kg	10/kg
O.D.V.	—	—
Germination (min.)	60%	60%
Moisture (max.)	7%	7%

SPINACH (*Spinacia oleracea* L.)

Spinach or Vilayati Palak belongs to family Chenopodiaceae is grown for its green tender leaves. It is strictly a cool season vegetable. It is sensitive to both low and high temperature.

Land Requirement

Spinach can be grown on the wide variety of soils. It is slightly tolerant to salinity. A pH range of 6–7 is ideal for spinach cultivation. The seed plot should be free from volunteer plants.

Fertilizers

Application of 50–70 kg nitrogen, 50 kg potash and 50 kg phosphorus/hectare is required. The entire dose of phosphorus and potash and one-fourth dose of nitrogen should be applied at the time of planting. The remaining quantity of nitrogen should be applied in two split doses as top dressing after first and second cutting.

Seed Sowing and Seed Rate

Spinach requires relatively higher seed rate of 35–45 kg/ha. as 50% of the plants turn out to be male with poor growth and such plants can not be removed till the time of blooming. In the plains the crop is sown from September–October and in hills from July to September. The distance between line to line is 30 cm and plant to plant distance may be maintained at 10–12 cm.

Irrigations

Being a shallow rooted crop it requires frequent irrigations at an interval of 5–6 days in summer and 8–10 days in winter.

Weed Control

One or two shallow hoeings are needed to keep the weeds under control. Cycloate @ 3–4 kg/ha. as pre-plant incorporation and Benthiocarb at 1.0 kg/ha or Methabenzthiazuron @ 1.0 kg/ha. as pre emergence spray successfully control the weeds of spinach.

Roguing

Off-type plants and diseased plants should be removed from seed plot time to time.

Isolation Distance

It is a wind cross-pollinated crop and the isolation distance should be kept at 1600 and 1000 meters from foundation and certified seeds, respectively.

Harvesting and Threshing

Usually for seed production purposes after 2–3 cuttings plants are left to produce stalks. The plants are cut after the seeds ripen and are left to dry. The seed do not shed from the fruits and hence can be left on plant to become fully mature. After proper drying threshing should be done and cleaned the seeds properly.

Field Standards	F/S	C/S
Off-type plants (max.)	0.10%	0.20%
Diseased plants (max.) (Seed borne)	0.10%	0.50%

Seed Standards	F/S	C/S
Pure seed (min.)	96%	96%
Inert matter (max.)	4%	4%
Other crop seeds (max.)	5/kg	10/kg
Total weed seeds (max.)	5/kg	10/kg
Germination (min.)	60%	60%
Moisture (max.)	9%	9%

11

SEED PRODUCTION OF SPICES

FENU GREEK (*Trigonella foenum-graceum* L.)

Fenu greek commonly known as Methi is a popular spice and leafy vegetable crop. Green leaves are rich source of minerals, protein, vit. A and C. The green leaves and tender stems are used as curried vegetable. Dried seeds are mainly used as spice for flavouring the food preparations.

The improved varieties of fenu greek are divided into two groups *viz.*, common methi and Kasuri Methi.

Land Requirement

Fenu greek is a winter season crop and requires moderately cool climate for its proper growth and development. Areas where heavy rains occur are unfit for its cultivation. It can be easily grown on all types of soils, however, well drained sandy loam soils with good fertility status having 6.0 to 7.0 pH are best. It is slightly tolerant to saline conditions. Apply Endosulfan powder @ 25 kg/ha to the soil before last ploughing to control the termites and other soil borne insects.

Fertilizers

Being a leguminous crop it is capable of fixing atmospheric nitrogen and requires lower dose of nitrogen. 30–40 kg nitrogen, 40 kg phosphorus and 40 kg potash/ha should be drilled in the soil at the time of sowing. Seed inoculation with *Rhizobium meliloti* should be done to increase to efficiency of nitrogen fixation.

Sowing Time

Fenu greek is sown from September to mid-March in the plains and April to October in hills. But for seed production last week of October to first week of November is the optimum sowing time in North Indian plains.

Seed Rates

25–30 kg/ha for common Methi

20–25 kg/ha for Kasuri Methi

Spacings

Line to line : 30 cm

Plant to plant : 5 to 7.5 cm.

Irrigation

First irrigation should be made just after sowing to facilitate better germination. When sowing is done in moist conditions, first irrigation is given at 30 days after sowing. In general, 4 to 5 irrigations are required.

Weed Control

Two weedings and hoeings at 25 and 50 days after sowing are sufficient to keep the crop free from weeds and maintain good aeration. Fluchloralin @ 0.75 kg/ha as pre-plant application + one hand weeding at 25 days after sowing is recommended to control weeds and resulted in better yield.

Roguing

Off-type plants as and when observed removed from the seed plot.

Plant Protection

Insect pests : Aphid (*Aphis craccivora*) suck plant sap from tender parts of flowers. This leads to yellowing of plants and resulted in shrivelling of grains and reduced yield of poor quality.

For control of Aphid spray Endosulfan 35 EC @ 0.07% or Dimethoate 30 EC @ 0.03% or Phosphamidon 85 SL @ 0.03% or Malathion 50 EC @ 0.03%. The spray may be repeated at 15 days interval if necessary. Mite (*Petrobia latens*) also suck the sap from young plant parts and inflorescence.

Control : Spray 0.02% emulsion of Ethion (50 EC.)

Diseases

Powdery Mildew : In the initial stage whitish powder may be seen on the leaves and under severe incidence whole plant is covered with powder.

Control : Dust sulphur powder @ 20–25 kg/ha or spray 0.2% solution of Wettable Sulphur or 0.1% Karathane or 0.05% of Calixin. This may be repeated at 15 days interval.

Downey Mildew : In the initial stage of disease a whitish growth on the lower surface of leaves appear. With the advancement of infection, the leaves start turning yellow and shedding and the plant growth is checked.

Control : Spray 0.2% solution of Indofil M–45 or Fylotan and repeat it after 15 days if necessary.

Leaf Spot : Numerous small, brown and circular spots are formed on the leaves. In severe incidence these spots increase in size.

Control : Spray Bordeaux mixture (5 : 5 : 50) or Blitox (0.3%) three times at 15 days interval.

Root Rot : Initially the plant shows yellowing and drying of leaves and later on whole plant may dry.

Control : Treat the seeds with Thiram or Captan @ 2–3 g/ kg seed.

Harvesting and Threshing

Seed crop matures in 150–160 days and Kasuri Methi in 160–170 days after sowing. The crop is harvested when the pods turn green to yellowish colour. Crop should be harvested at this stage other wise delay in harvesting leads to shattering of the seeds.

Field Standards	F/S	C/S
Off-type plants (max.)	0.10	0.20
Objectionable weeds (max.)	0.01	0.02

Seed Standards	F/S	C/S
Pure seed (min.)	98%	98%
Inert matter (max.)	2%	2%
Other crop seeds (max.)	10/kg	20/kg
Total weed seeds (max.)	10/kg	20/kg
Objectionable weed seeds (max.)	2/kg	5/kg
O.D.V. (max.)	10/kg	20/kg
Germination (min.)	70%	70%
Moisture (max.)	8%	8%

CORRIANDER (*Corriandrum sativum* L.)

Corriander is an annual herb and grown for both green leaves and dried seeds. The fruits are considered as carminative, diuretic, anti bilious, refrigerant and aphrodisiac. It is a tropical crop and grown in a cool season crop. It is susceptible to frost, particularly during flowering and grain formation stages. Cloudy weather during flowering and fruiting is the main cause of spread of pest and diseases. Moderately cool and dry weather during grain formation stage facilitates higher yield and better quality seeds.

Land Requirement

Corriander can be grown on almost all type of soils having sufficient organic matter. However, loam soils are most suited. It is quite sensitive to saline and alkaline soils.

Fertilizers

Apply 20 kg nitrogen, 30 kg phosphorus and 20 kg potash per hectare as basal dose. Besides this, 40 kg nitrogen/ha is applied into two equal split doses at 30 and 60 days after sowing.

Sowing Time

Corriander can be grown throughout the year for leaf purpose. For seed purpose it is sown from October to November at 30 × 15 cm space.

Seed Rate

10–15 kg/ha is sufficient. Soaking of the seeds in water for 12–24 hours before sowing enhance seed germination. The seed split into two halves by rubbing and treat them with Thiram @ 2g/kg seed.

Irrigation

First irrigation should be given 3 days after sowing and there after at 10–15 days interval depending upon the availability of moisture in the soil. In general 5 irrigations are applied.

Weed Control

It is very important to control the weeds in corriander especially at the initial stage as growth of weeds is much faster than the initial growth of the crop. The first weeding and hoeing along with thinning of plants should be carried out at 25 days after sowing and again at 50 days after sowing.

Roguing

All the Off-types plants and diseased plants should be rogued out from the seed plot as and when observed.

Plant Protection

Insect-pests : Aphid (*Hyadaphis corrianderi*) suck plant sap from tender parts and flowers. This leads to yellowing of plants and resulted in shrivelling of grains.

Control : Spray Endosulfan 35 EC @ 0.07% or Dimethoate 30 EC @ 0.03% or Phosphamidon 85 SL @ 0.03% or Malathion 50 EC @ 0.03%. The spray may be repeated at 15 days interval if necessary.

Mite also suck the sap from young plant parts and inflorescence.

Control : Spray 0.02% emulsion of Ethion (50 EC).

Diseases

Powdery Mildew : A white powdery mass appears on the leaves and twigs of plants. At later stages, the whole plant is covered with the whitish powder. The infected plants do not produce seeds.

Control : Dust sulphur powder @ 20–25 kg/ha. Spray 0.2% solution of wettable sulphur or 0.1% Karathane at 15 days interval.

Blight : The disease appears in the form of dark brown spots on the stem and leaves. Later on infected leaves appear blighted.

Control : Spray 0.1% solution of Bavistin or 0.2% solution of Indofil M–45.

Stem gall : Blisters appear on the leaves and stem which deforms the seeds. Humid conditions are favourable.

Control : Treat the seeds with Agrosan GN @ 2g/kg seed or Thiram and Bavistin in the ratio of 1 : 1 @ 2g/kg seed. Spray 0.1% solution of Bavistin and repeat two to three times at an interval of 15–20 days.

Wilt : The disease attacks the roots of the plants which wilt and slowly dry up.

Control : Treat the seeds with Thiram and Bavistin in the ratio of 1 : 1 @ 2g/kg seed.

Harvesting and Threshing

Corriander matures in about 90–145 days depending upon the variety and growing conditions. Change from green to yellow colour of grains is an indication of maturity. Harvesting should be done at a stage when 50 per cent grains turn yellow to obtain good lusture of the grains.

ONION (*Allium cepa* L.)

Onion is a biennial crop and takes two full seasons to produce seeds. In the first year bulbs are formed and in the second year stalks develop and seed is produced.

Land Requirement

Onion can be grown in different types of soils. The pH of the plot should vary from 6–8. Clay soils are not suitable for onion cultivation. Field selected should not have been under the same crop in the previous seasons.

Isolation Requirement

It is highly cross-pollinated crop.

	F/S	C/S
For mother bulb	5 m	5 m
For seed production	1000 m	500 m

Seed Season and Sowing

Onion can be sown in all the three season (June-July, September-October, January-February). Six to eight kg seed is required for 1 hectare area.

Methods of Seed Production

For onion seed production two methods are followed :

1. *Seed to Seed Method :* In this method bulb crop is let to over winter in the field so as to produce seed in the following season. This is also called *"in situ"* method.

2. *Bulb to Seed Method :* The bulb produced in the previous season are lifted, selected and stored. In the second year replanted the bulbs for seed production.

In seed to seed method nursery is planted in May-June and transplanting is done in the month of July-August. Bolting started in January-February and seeds are ready in middle of May.

However, in bulbs to seed method, bulbs are produced in the previous season and replanted to produce seed in the second year. Nursery is sown in October–November, transplanted in December-January and the bulbs are lifted by the end of May. True to type bulbs are selected and stored upto September–October and then planted in field.

Fertilizer Application

A total fertilizer dose is required 100–125 kg nitrogen, 50 kg phosphorus and 75 kg potash per hectare. Half nitrogen and entire dose of phosphorus and potash should be applied as basal. The remaining half dose of nitrogen is to be given after 6 weeks.

Transplanting

Transplanting should be done when seedlings are 6–8 weeks old. The spacing between row to row and plant to plant is 15 and 10 cm, respectively.

Irrigation and Interculture

Irrigate the crop once in 8–10 days depending upon seasonal conditions. Light hoeings and 2–3 weedings are required to control the weeds.

Plant Protection Measures

In onion the major insect is onion thrips and purple blotch is important disease.

Onion thrips can be controlled by spraying of Malathian @ 1 ml/lit. of water. If required spray may be repeated. For the control of purple blotch, spray the crop regularly with Mancozeb or 30g Difolton in 10 lit. of water. Repeat the spray after 15 days.

Roguing

Depending upon the foliage colour, plant type or late maturing Off-types as thick-necks, bottle-necks which do not conform to the varietal character.

Harvesting and Curing of Bulbs

Well matured bulbs should be harvested. Maturity is indicated by dropping of the tops just above the bulb, while the leaves are still green. After harvesting, the bulbs should be topped leaving an half inch neck. Before storage, a thorough selection and curing of bulbs should be done. Remove injured and rotten bulbs including pre-mature bulbs.

Planting of Bulbs for Seed Production

The size of the bulb for planting purpose should be medium ranging from 2.5–3.0 cm in diameter. The best sowing time is second week of October to third week of October. The bulbs are planted at a distance of 45 × 30 cm. About 20–25 quintal bulbs are required for planting one hectare. Irrigate the field just after planting. The bulbs are planted 8–10 cm deep in the soil.

Harvesting and Threshing

All heads do not mature simultaneously, therefore harvesting is done in 2–3 instalments. When the seed inside the capsule become black the umbel is cut, keeping a small portion of the stalk attached. The umbels are dried in the sun and thresh by hand. After proper cleaning and drying the seed should be stored properly.

Seed Standards	F/S	C/S
Pure seed (min.)	98%	98%
Inert matter (max.)	2%	2%
Other crop seeds (max.)	5/kg	10/kg
Weed seeds (max.)	5/kg	10/kg
Germination (min.)	70%	70%
Off-types (max.)	0.10%	0.20%

12

SEED PRODUCTION OF COLE CROPS

CAULIFLOWER

Cauliflower requires a cool, moist climate for seed production. As for as possible it is better to avoid light soil as they form small and loose curds. The seed of Asiatic cauliflower is successfully produced in North-eastern plains. The temperate varieties flourish well in Solan area of H.P.

Land Requirement

Land to be used for seed production should be free of volunteer plants. The soil of selected field should be fertile with a pH value 5.5 and well drained.

Isolation

Cauliflower is mainly cross-pollinated crop. Pollination is mainly done by bees.

Foundation seed : 1600 meters

Certified seed : 1000 meters.

Methods of Seed Production

1. *In situ* method (seed to seed method)

2. Transplanting method (head to seed method)

For seed production, seed to seed method is recommended. In this method, the crop is allowed to complete its vegetative phase, bolt flower and produce seed in the same field where seedlings are transplanted.

Seed, Season and Sowing

400–600 g for planting one hectare of land. Nursery may be sown in May-June for early varieties in plains and mid season varieties in late June-July in plains and late temperate varieties in late August in hills.

Transplanting

The seedlings of 12–15 cm height or 5–6 leaf stage are fit for transplanting. The seedlings should be transplanted during evening hours. Light irrigation should be provided after transplanting.

Spacing

Early and mid season – 60 × 45 cm.

Late – 60 × 60 cm.

Fertilizers

The Cauliflower crop requires heavy manuring as it removes large quantities of major nutrients from the soil. For best results apply 50–60 tonnes of FYM/ha at the time of field preparation. Apart from this F.Y.M. apply 100–120 kg nitrogen, 70–80 kg phosphorus and 40–50 kg potash per hectare. Half dose of nitrogen and entire dose of phosphorus and potash should be supplied as basal. Remaining half dose of nitrogen should be applied as top dressing in two split doses.

Sometimes crop showed deficiency of boron and molybednum. Apply 10–15 kg borax per hectare to the soil or two sprays with 0.3 per cent borax on the seedlings may correct the boron deficiency. For molybednum deficiency apply 1–1.5 kg/ha of sodium molybdate.

Irrigation and Interculture

Irrigate the crop as and when requirement of the crop depending upon climatic conditions. Frequent shallow inter culture operations are needed for good crop.

Plant Protection

The major pests are aphid, leaf webber, semilooper, diamond black moth, head borer, cabbage bug and cut worm. The major diseases are bacterial black rot, collor rot, blight, white rust, damping off, downy mildew and mosaic. Spray the nursery bed with 5 ml phosphomidon with 30 g copper oxychloride dissolve in 100 lit. of water, 2 weeks and 4 weeks after sowing. Spray the same mixture 2 weeks and 4 weeks after transplanting. Again the crop should be sprayed with 20 ml, malathion with 60 g copper oxychloride in 10 lit. of water, 6 and 8 weeks after transplanting.

Indiscriminate use of insecticides should be avoided to protect the bees during flowering which are pollinators.

For the control of Damping off drench the nursery with 150 g of captan in 100 lit. of water to cover 2000 sqm. area before sowing. Whenever, black rot infection is noticed, spray the crop with 100 mg Tetracycline hydroxide dissolved in 10 lit. of water.

Scooping

Scooping the central portion of curd when it is fully formed helps in the early emergence of the flower stalks.

Roguing

Selection of curds is done when the curds are well developed. Off-type plants, and those forming poor curds, should be removed at this stage.

Subsequent roguings for Off-types and diseased plants affected by black-leg, black rot, leaf spot and phyllody should be done from time to time.

Harvesting and Threshing

Harvesting can be done when siliquae turn brown. Too ripe siliquae dehisce. Seed should not crush or split when rubbed between the hands.

Generally the early plants are harvested first, when about 60–70 per cent of the siliquae turns brown and rest of the crop changes to yellowish–brown. After harvesting it is piled up for curing. After 4 to 5 days it is turned upside down and allowed to cure for another 4–5 days in the same way. It is then threshed with sticks. After proper cleaning and drying seed should be stored.

Field Standards	F/S	C/S
Off-type plants (max.)	0.10%	0.20%
Diseased Plants (max.)	0.10%	0.50%

Seed Standards	F/S	C/S
Pure seed (min.)	98%	98%
Inert matter (max.)	2%	2%
Other crop seeds (max.)	5/kg	10/kg
Weed seeds (max.)	5/kg	10/kg
Germination (min.)	65%	65%
Moisture (max.)	7%	7%

CABBAGE (*Brassica oleracea* Var. Capitata L.)

Cabbage belongs to the family Cruciferae. It is a herbaceous annual for vegetable, where as for seed production it is biennial. It has bisexual flowers. It is a temperate crop which does not produce seeds in the plains.

Land Requirement

Land to be used for seed production should be free of volunteer plants. The soil of selected field should be fertile.

Isolation

Cabbage is mainly cross–pollinated crop. Pollination is mainly done by bees.

Foundation seed : 1600 meters

Certified seed : 1000 meters

Methods of Seed Production

Being biennial, the cabbage requires two seasons to produce seed. In the first season the heads are produced, and in the following season seed is produced. The seed crop can be left *in situ* or transplanted during autumn. *In situ* method is usually followed for seed production.

In the *in situ* method, the crop is allowed to over-winter and produce seed in their original position, that is, where they are first planted in the seedling stage. In the transplanting method, the mature plants are uprooted. After removing whorls the plants are immediately reset in a well prepared new field, in such a way that the whole stem below the head goes underground with head resting just above the surface.

Stump Method

The matured head, which are true to type, are cut just below the base, keeping the stem with outer whorl intact. The head removed are marketed and the remaining portion which is called stump is either left *in situ* or transplanted in the second season. After dormancy is broken, the bud sprouts from axil of all the leaves and leaf sears. The flowering shoot requires heavy staking to avoid its breaking during field operations.

Stump with Central Core Method

After the selection of the head which are true to type, they are chopped on all sides with downward perpendicular cuts in such a way that the central core is not damaged. The shoots arising from the main stem are not decumbent, hence heavy staking is not required.

Head Intact Method

The heads which are true to type are kept intact and only a cross cut is given to facilitate the emergence of the stalk.

Seed, Season and Sowing

Seed rate : 500–700 g/hectare

Nursery may be raised in June-July for early varieties in mid hills and late varieties in May-June in mid and high hills.

Transplanting

The seedlings of 5–6 leaf stage are fit for transplanting. The seedlings are to be planted during evening with a spacing of 45×45 cm (early) and 60×60 cm (late). Just after transplanting a light irrigation must be provided.

Fertilizers

Apply 50–60 tonnes FYM/hectare at the time of field preparation. In addition to this apply 100–150 kg nitrogen, 40–50 kg phosphorus and 30–40 kg potash per hectare. Entire dose of phosphorus, potash and 1/3 of nitrogen should be given as basal. Remaining quantity of nitrogen should be applied in three split doses. Boron deficiency may be corrected by one or two sprayings of 0.3% Borax and Molybednum deficiency may be corrected by the application of line or 1-1.5 kg sodium molybdate/hect.

Irrigation

Irrigate the crop as and when required.

Plant Protection : Same as in cauliflower.

Roguing

The first roguing is done at the time of handling the mature heads. All Off-type plants/diseased plants and undesirable types are removed.

Second roguing is done before the heads start bursting. The loose, leaved poorly heading plants, and those having a long stem and heavy frame, must be rogued out. Subsequent roguing for Off-type, diseased plants should be done time to time.

Harvesting and Threshing

To avoid shattering of seeds, the whole crop is harvested

in two-three lots with sickles. When crop changes to yellowish-brown harvest the crop and stake for curing. After 4-5 days crop is threshed, clean and dried.

Field and Seed Standards	F/S	C/S
Pure seed (min.)	98%	98%
Inert matter (max.)	2%	2%
Other crop seeds (max.)	5/kg.	10/kg.
Weed seeds (max.)	5/kg.	10/kg.
Germination (min.)	65%	65%
Moisture (max.)	7%	7%
Off-type (max.)	0.10%	0.20%
Diseased plants (max.)	0.10%	0.50%

13

SEED PRODUCTION TECHNOLOGY OF FLOWER CROPS

Production of seeds of floricultural crops is a lucrative trade as the return per capita is much more than any other branch of agriculture.

The four major factors *viz.*, environmental, agronomic, cultural and post-harvest operational factors determine the quality. Among the environmental factors light, which affect the photosynthetic activity of plants is important in floriculture. Thus, they are short day, long day or day neutral plants according to their specific light requirement.

Steps to Produce Good Quality Seeds

At Production Stage : Proper fertilization, adequate watering, sufficient isolation, adequate roguing, timely harvest and care after harvest are the important steps during production stage. For maintenance of genetic purity of the crop is mainly due to maintenance of proper isolation distance and conducting the roguing operation at appropriate stages.

Isolation

The seed crop must be isolated from other nearby field of other varieties of the same crop and other contaminating crops.

Roguing

Adequate and timely roguing is extremely important in seed production. The rogues which differ from the normal plant population are to be removed at the earliest possible date before flowering. It is wise to remove whole plant and not just the flower head.

1. Roguing at vegetative/pre-flowering stage.

2. At flowering stage.

3. At maturity stage

1. **Vegetative/pre-flowering stage :** This stage, in cross pollinated crops is extremely important to avoid genetic contamination. The plant obviously differing in height, colour, vegetation, leaf size, shape and orientation or any other easily distinguishable characters such as malformed/diseased plants should be removed completely.

Flowering Stage : The roguing at flowering stage is equally important, perhaps even more important than at the vegetative stage. The undesirable plants not distinguishable earlier should be removed soon after the identification (observed) in order to avoid genetic contamination. In this stage identify the undesirable plants based on flower characters.

Maturity Stage : Roguing of Off-type plants not distinguishable earlier. Roguing of the harvested fruits, tubers, roots, corms may be necessary to remove off-textured/Off-coloured fruits.

Harvesting : An ideal time for a given seed crop is the point beyond which losses will be greater. The seed moisture is also an indication of crop fitness for harvesting. The small seeded flowers are harvested manually. To avoid shuttering harvest the crop before opening of seed head in the field.

Threshing : The stage of maturity and moisture content determines the efficiency of threshing. For winnowing cleaning and grading the traditional and mechanical methods are employed.

Processing, Drying and Storage : Damaged seeds, mixture and proper seed treatment and suitable packaging should be done at proper moisture level.

Rose (Rose sp.) : Seed and vegetative methods like cutting, layering, budding and grafting.

Season	:	October to early April
Spacing	:	60 × 30 cm
Fertilizer	:	180 : 180 : 190 kg NPK/ha.
Yield	:	15-30 flower/plant

Chrysanthemum (*Chrysanthemum morifolium*)

Propagation : Seed and vegetative methods like multiplication through sucker, cutting, air layering, micro propagation.

Season	:	May to middle August.
Spacing	:	12-16 m^2
Fertilizer	:	204 : 68 : 408 kg NPK/ha.
Yield	:	150 000 to 175 000 flowers/hectare

Gladiolus (*Gladiolus hybridus*)

Propagation : Seed and vegetative methods like cormels and corms. Divisions of corms and micro-propagation.

Season	:	September to November
Spacing	:	25-30 m^2
Fertilizer	:	8-15 tonns/ha. FYM
		80-150 kg Nitrogen
		40-60 kg P$_2$O$_5$
		100-200 kg K$_2$O/hectare and 30-40 kg mg/ha.
Yield	:	200000-300000/ha.

Carnation (*Danthus aryophyllus*)

Propagation : Seed and vegetative methods like cuttings and micro-propagation.

Season	:	September to April
Spacing	:	25-32 plants/sqm.
Fertilizers	:	317-400 : 54-73 : 680-726 kg/NPK/ha., 100-150 kg Mg/ha.

Tuberose (*Polianthes tuberosa*)

Propagation	:	Seed and vegetative part *i.e.,* bulb.
Season	:	February-March
Spacing	:	20 × 20 cm
Fertilizers	:	150-200 : 80-100 : 150 kg NPK/ha.
Yield	:	5.6-9.6 tonnes/ha.

Gerbera (*Gerbera jamesoni*)

Propagation	:	Seed and division of clumps/ Suckers and micro-propagation
Spacing	:	8-10 plants/m^2
Yield	:	175/m^2/year.

Table 1 : Seed Technological Requirements for Important Crops

Common Name	Botanical Name	Propagating Material	Spacing	Fertilizers (NPK kg/ha.)	Season	Isolation distance (meters)
China aster	*Callistephus Chinensis*	Seed	30 × 20 cm	40 : 40 : 0	August-September	400
Ageratum	*Ageratum conyzoides*	Seed	30 × 20 cm	40 : 40 : 0	March-April	400
Alyssum	*Alyssum maritimum*	Seed	30 × 20 cm	40 : 40 : 0	October-November	400
Calenchula	*Calenchula officinalis*	Seed	30 × 30 cm	40 : 40 : 0	February-March	400
Dahlia	*Dahlia variabilis*	Seed	90 × 45 cm	40 : 40 : 0	September-October	
		Cutting	75 × 45 cm	40 : 40 : 0	March-May	400
Gerannium	*Pelargonium hortorum*	Seed				
Marigold	*Tagetus erecta*	Seed	40 × 30 cm	40 : 40 : 0	May-June	400
	T. petula	Seed	25 × 15 cm	40 : 40 : 0	August-October	
Pansy	*Viola tricolo*	Seed	25 × 15 cm	40 : 40 : 0	September-October	200
Petunia	*Petunia hybrida L.*	Seed	30 × 30 cm	40 : 40 : 0	August-March	100
Snapdragan	*Antirrhinum majus L.*	Seed	40 × 20 cm	40 : 40 : 0	September-October	—
Sweet pea	*Lathyrus odoratus L.*	Seed	30 × 15 cm	40 : 40 : 0	February-March	100

14

INTERNATIONAL SEED ORGANIZATIONS

Development of Seed Technology has been taken place internationally primarily due to seed exchange and seed trade between different countries. For seed production most multinational companies are shifting areas of seed production from home base to other areas in the world, more quantities of seed are entering in the international trade. This trade is primarily depending upon the assurance of quality seed from exporting/producing countries to seed importing countries. In order to provide certification, seed testing and other quality control some important international organizations are played a vital role.

ISTA (International Seed Testing Association)

In the early stages of seed testing, the need was felt for the cooperation between the Seed Testing Laboratories all over the world and as a consequence of this, first international meeting on seed testing was held in 1870.

Subsequently the International Seed Testing Association came into existence in 1924 for ensuring uniformity in seed testing and to determine different seed quality testing methods. ISTA rules are used in seed testing by its member and non-member countries of the world. It has a system of International and National Referee testing through which the participating stations can appreciate their short comings. On the behalf of international seed trade, ISTA distribute standardized international certification, *i.e.*, orange/green and blue analysis certificates. Orange certificates are used when the official sampling and final testing occur in the same country. Green certification used when a lot sampled in one country and analysed in other country. Blue certificate is issued in respect of seed samples submitted by an agency for analysis to the ISTA accredited station. Here the sampling and sealing is not the responsibility of the accredited seed testing laboratory.

To become member of ISTA, it is necessary that one is engaged in seed activity must be designated by their respective government to participate in the Association on behalf of the government. India became an accredited member of ISTA in 1962.

The secretariate is being changed every 10th year among the ISTA member laboratories. Previously it was in Denmark, Netherlands and Norway and now it is located in Zurich, Switzerland.

For communication, ISTA has recognised three languages, English, French and German. The first set of Rules was approved in 1931 in Wageningen.

Objectives

1. Uniform application of procedure for evaluation of seed moving in international trade.

2. Activity promote research in all areas of Seed Science and Technology.

3. To participate in conferences, and training courses aimed at furthering the above objective.

4. To establish and maintain liason with other organizations having interest in seed.

5. Published an International Journal, "Seed Science and Technology".

AOSCA (Association of Official Seed Certifying Agency)

In 1919 International Crop Improvement Association (ICIA) was formed and enunciated the following fundamental concept of seed certification :

1. Pedigree of all certified seed crops must be based on lineage.

2. The integrity of certified seed growers must be recognised.

3. Field inspection must be made by qualified inspector.

4. Verification trials to establish identification and usefulness of the varieties and strains certified must be conducted.

5. Keeping of proper records to establish and maintain satisfactory pedigree of seed stock.

6. Establishment of standards for purity and germination.

7. Principle of sealing seeds to protect both grower and purchaser must be approved.

8. Species of farm weeds which would be included within the meaning of abnoxious weeds as listed by ICIA must be defined.

9. There must be standardization of nomenclature used, describing the class of pedigreed seed.

In early forties, ICIA prepared a set of minimum seed standards for certification purpose and published them in 1945, These standards were revised from time to time according to the need of the certifying agencies of USA, Canada and some other countries. In 1969, ICIA changed its name to the Association of Official Seed Certifying Agency (AOSCA).

OECD (Organization for Economic Cooperation and Development)

It was developed after the second world was in Europe. It is an inter-governmental organization and having 27 countries as members. The main objectives of this organization are :

1. Allow seed produced under the rules and directions of the scheme can be moved in international trade.

2. Extend the principles, procedures and standards of seed certification to other countries of the world for

using the OECD model for their domestic certification scheme.

3. Every year a list is completed of varieties eligible for certification.

4. Field tests are conducted to ascertain that the characters of varieties remain unchanged in the process of multiplication and enable the trueness to variety and varietal purity of new lots.

FIS (International Seed Trade Federation)

It is the global organization established in 1924. It is a non-profit group of national association as well as individual seed companies. It has 60 member countries. The mission of FIS is :

1. To promote the interests of seed industry.

2. To develop and facilitate the free movements of seed with fair and reasonable regulations.

3. To serve farmers, growers and consumers and protect intellectual property.

4. Encourages the use of modern technologies in high quality seeds for developing sustainable agriculture for the production of food and fibre in a healthy environment.

UPOV (International Union for the Protection of New Varieties of Plants)

The purpose of the UPOV is to ensure that the member states acknowledge the achievements of breeders of new plant varieties, by making available to them an exclusive property right to their varieties, on the basis of a set of uniform and clearly defined principles. To be eligible for protection, varieties have to be :

1. Distinct from existing, commonly known varieties.

2. Sufficiently homogenous (uniform).

3. Stable.

New, in the sense that they must not have been commercialized prior to certain dates established by reference to the date of application for protection. Protection is provided as incentive to the development of agriculture, horticulture and forestry and to safeguard the interests of plant breeders. The 1978 convention of UPOV farmers and plant breeders are exempted for use of protected varieties, but in the 1991 convention no one is exempted and India is signatory to it.

APSA (Asia and Pacific Seed Association)

The aim of APSA is to improve production and trade in the Asia-Pacific region (including Newzealand, Australia and the USA), as well as the other Asian countries; of quality seed and planting material of agricultural and horticultural crops. Functioning as a regional form, the Association :

1. Represents the interests of members to governments.

2. Encourages collaboration among seed enterprises in the region, strengthening links with International Organizations.

3. Complete and disseminates information on technical regulatory and market issues.

4. Association organizing training and cultivar testing programmes.

ASSINEL (International Association of Plant Breeders for the Protection of Plant Varieties)

This is a sister organization of FIS. Both FIS and ASSINEL take a strong position on farm-saved seed *i.e.*, seed saved by a farmer to be sown the following year, they call it illegal. The mission of ASSINEL is to :

1. Increase recognition of the importance and value of the plant breeder's contribution to world agriculture and horticulture.

2. Represent at international level, to protect the IPR (Intellectual Property Right) of plant breeders.

FAO (Food and Agricultural Organization)

This is a system for use in countries where there are insufficient resources for a fully developed seed quality control scheme such as seed certification. Its aims are :

1. To ensure the resources available in a country are used to the best advantage to the farmers to purchase quality seed.

2. To encourage and assist the development of technical expertise in a seed industry.

AOSA (Association of Official Seed Analyst)

The membership of AOSA is open to government seed laboratories in the U.S. and Canada; other can participate as associate non-voting members. The objective of this organization is : "Uniformity in methods of seed testing".

AOSA is further increasing its cooperation with the SCST (Society of Commercial Seed Technologist) which represents independent seed company with private analysts.

International Organizations

There are twelve International Institutes out of which ten institutes are directly or indirectly involved in crop improvement work and supplement in national crop improvement efforts. These institutes are situated in tropical countries. The main objective of these institutes is to increase agricultural production of tropical countries through applied research coupled with extension and educational activities. The functioning of these institutes is supported and supervised by the CGIAR (Consultative Group for International Agricultural Research).

CGIAR was established in 1971 by the joint efforts of FAO (Food and Agricultural Organization), bank and UNDP (United Nations Development Programme).

IRRI (International Rice Research Institute)

This institute was initiated in 1960 jointly by Ford

Foundation and Rockefeller Foundation at Manila, Phillipines. This institute is concerned with the improvement of rice genotypes for tropical countries. In addition, it maintains a huge collection of rice germplasms approximately more than 42000 gene pools. This institute has a sophisticated long term storage facility for rice germplasm.

CIMMYT (International Centre for Maize and Wheat Improvement)

It was established in 1966 by the joint efforts of Ford Foundation and Rockefeller Foundation at Mexico. The primary responsibility of CIMMYT is the improvement of Maize and Wheat. The improvement for other cereals (*viz.*, triticale, barley and jawar) is also carriedout. The major contribution of CIMMYT to our country is development of Mexican semi dwarf wheat and open-pollinated varieties of Maize. It maintains a huge collection and maintenance of maize and wheat germplasms.

CIAT (International Centre for Tropical Agriculture)

This institute was established in 1967 with joint efforts of Ford Foundation and Rookefeller Foundation. The improvement on cassava and beans is the primary responsibility of CIAT. This institute also works on rice and maize improvement with the collaboration of CIMMYT and IRRI. The institute also carries out research on pasture management and farming system. It has a good collection of beans, forage grasses and pasture legumes.

IITA (International Institute of Tropical Agriculture)

This institute was established in 1968 with a objective for the improvement of grains legumes, root and tuber crops. Research work on cropping system is the major responsibility of this institute. It is basically concerned with the problems of African countries. It has a huge collection of cowpea germplasm.

WARDA (West African Rice Development Association)

This institute is established in 1971 by west African Government. It looks rice improvement programme with collaboration of IRRI and IITA. The major emphasis is carried out on rice research for African countries.

CIP (International Centre for Potato)

This institute is established in 1971 by CGIAR at Lema, Peru. The major responsibility of institute is the improvement on potato for tropical countries. The institute has a large number of potato germplasms.

ICRISAT (International Crop Research Institute for Semi-Arid Tropics)

This institute was established in 1972 by CGIAR at Patancheru, Andhra Pradesh. This is a non-profit scientific educational institute. The institute is mainly concerned with the improvement work on Sorghum, pearlmillet, chickpea, pigeonpea and groundnut. The institute also, conducted research on dryland farming system. This institute has a huge collection of germplasms of sorghum, pearl millet, chickpea, pigeonpea and groundnut.

ICARDA (International Centre for Agricultural Research on Dry Areas)

This was established in 1977 by CGIAR at Aleppo, Syria. It undertakes research relevant to the needs of developing countries and specifically for the agricultural systems in West Asia and North Africa. The overall objective of the Centre is to contribute towards increased agricultural productivity, there by increasing the availability of food in both rural and urban areas, and thus improving the economic and social well-being of people.

It is a world centre for the improvement of barley, lentil and fababean and a regional centre for the improvement of wheat, chickpea, farming systems, pasture and forage crops and live stock. Training agricultural researchers from developing countries is an important component of ICARDA's activity.

IPGRI (International Plant Genetic Resources Institute)

This institute was established in 1994. Earlier it was known as IBPGR (International Bureau of Plant Genetic Resources) was established in 1974 by CGIAR. The aim of this institute is to coordinate and promote the conservation of plant genetic resources. It also supports several crop specific local programmes, *e.g.*, rice collection at IRRI, and also provides information on plant genetic resources.

15

PLANT VARIETY PROTECTION AND FARMERS RIGHTS ACT (PVP AND FR ACT, 2001)

The increased food production ushered by in so called Green Revolution in late 60's and early 70's has been to the extent of nearly four times taking the food production of 1950 as bench mark. This increased food production could become possible with the advent of new breed of crop varieties particularly the cereals like rice and wheat as well as by exploitation of heterosis in some other crops. The breeding of new crop varieties by using approaches like hybridization and selection was considered as natural and obvious discoveries with low investment. Further, in this both the product and processes can rarely be copied. Hence it did not warrant any protection or patenting.

However, some developments at global level have forced us to think of protecting the plant varieties and also the farmers rights. The first development is the one concerning WTO (World Trade Organization). WTO was established in 1995, which determines the global rules of trade between the nations. India is a signatory to the WTO and to TRIPS (Trade Related Intellectual Property Rights), where in, it is necessary as per article 27, 3(b) (*ii*) to protect new plant varieties through patents or through a *sui generis* system or by combination of both. The second development is the impact of modern technology in plant breeding. The inventions in biotechnology are believed to require extensive technical knowledge and expertise. Further, they require substantial investments contrary to the conventional methods of plants breeding. The processes and products can be copied which warrants protection. The technological change will not be possible without substantial investment in R & D. The Govt. alone will not be in a position to pump in the capital needed for bringing about the desired technological shift. All these led India to adopt the "Plant Variety Protection" and it opted to evolve a *sui generis system* for PVP and FR Act. Indian Patents Act of 1970, did not cover the living organisms or methods of agriculture. With signing of WTO by India, extensive discussions were held between 1995-2000 over *sui generis* PVP bill. Ultimate, Lok Sabha passed a PVP-FR Act on August 9, 2001.

Objectives

1. To recognise and protects the rights of the farmers, for their contribution in making available genetic resources for the development of new varieties of plants.

2. To stimulate investment for research, development of new plant varieties and to promote agricultural research by both the public and private sector.

3. To facilitate growth of seed industries and to ensure availability of high quality seeds and planting material to the farmers.

4. To establish National Gene Fund and introduce a concept called "benefit sharing" where by the efforts of farmers, who contributed to the development of new plant varieties, are duly recognised.

Salient Features

India, by enacting PVP Act, has declared that it will adopt a *sui generis* type of protection for protecting new plant varieties. The Act is presented in 11 Chapters comprising 9 Sections. Chapter 2, comprising of Sections 3-13 about the PVP Authority, its functions, powers, composition and the maintenance of National Register of Plant varieties. Chapter 1, Section 2 (h) defines benefit sharing. The effect of this concept is that as and when a new plant variety is developed from out of the genetic material preserved by a special or particular farming community, members of the community can demand from the National Plant Varieties and Farmer's Right Authority, the royalities that are due from the breeders. However, foreign nationals have been prohibited from invoking the concept of benefit sharing. The denial of rights to foreign nationals is a violation of the principle of right to national treatment, which is the corner stone of Paris Convention on the Protection of Industrial Property 1882. In addition, under the traditional patent law, existing knowledge or prior art is free for all. Any one is quite free to take advantage of the same. Now the law says that some in the farming community can demand that breeder pay royalty for the genetic material (part of the prior art) without having made any contribution whatever, except belonging to a particular community. Whether it will be legitimate to compensate a few in the community or how to determine the beneficiary for royalties opens up a further set of difficult questions and the authority constituted under the Act is an unknown angle in comparison with the known devils functioning under other legislations.

Section 2(3) defines breeder to mean a person or group of persons or a farmer or group of farmers or any institution, which has spread, evolved or developed any new variety. The same Act in Section 2(*k*) defines a farmer to mean any person who

1. Cultivates crops by cultivating the land himself.

2. Cultivates crops by directly supervising the land.

Chapter III of the Act speaks about the Registration of Plant Varieties. Section 16 says that for a new variety to be entitled to protections, it must comply with the following criteria:

1. **Novelty :** A variety should not have been commercially exploited for more than one year before grant of PBR protection.

2. **Distinctness :** The new variety must be distinguishable from other varieties by one or more identifiable morphological, physiological or other characteristics.

3. **Uniformity :** The new variety must be uniform in appearance under the specified environment of its adaptation.

4. **Stability :** That is, such essential characteristics remain unchanged after repeated propagations under the specified environment.

Section 15 of this Act makes it clear that the protection will not be available in certain situations that include the employment of the terminator technology or the plants that are designed not to produce seeds. Section 19 stipulates that the applicant shall make available, along with the application, prescribed quantity of seeds for conducting tests of distinctness, uniformity and stability.

Chapter IV of the Act comprising Sections 24-32, describes the duration and effect of registration besides explaining the concept of benefit sharing. Section 24 provides the following life term for different plant varieties :

1. In the case of trees and vines, 18 years from the date of registration of the variety.

2. In the case of extinct varieties 15 years from the date of notification of that variety by the central government under Section 5 of the Seeds Act, 1966.

3. In the other cases, 15 years from the date of registration of the variety.

AS PER SECTION 14 OF THE ACT

A Protection is available for those general species as have been notified as varieties entitled for protection under Section 29(2) of the Act.

B Protection is available for extant varieties which are notified under Section 5 of the Seeds Act 1966 or farmers variety or varieties about which there is common knowledge or any other variety which is in public domain.

C   Farmer's variety which has been defined under Section 2(*i*) means a variety which has been traditionally cultivated and evolved by the farmer in the fields or a variety which is a wild relative or land race of a variety about which the farmers posses common knowledge.

The Act also recognizes the freedom of scientific research. Section 30 of the Act prevents.

A   The use of any variety registered under this Act, by any person using such variety for conducting experiment or research.

B   The use of variety by any person as an initial source of variety for the purpose of developing other varieties.

Provided that the authorization of the breeder of a registered variety is required where the repeated use of such variety as a parental line is necessary for commercial production of such other newly developed variety. Chapter V of the Act comprising Sections 33-38 lays down provisions for surrender, revocation, rectification and correction of Register of Plant Varieties. By and large these are very similar to the provisions prevailing under the Law of Patents.

Chapter VI entitled, as "Farmer's Right" comprises Section 39-46 :

(*i*)   These right include the right to save, use, sow, resow, exchange, share or sell his farm produce including seed of a variety protected under this Act. However, this right will not entitle him to sell branded seed of a variety.

(*ii*)   Farming communities who have contributed to the conservation of the genetic material used in the evolution of a new variety are entitled to royalties as may be determined by the PVP authority. The Act also contemplates the creation of "National Gene Fund" and stipulates that the dues from the breeder may be recovered as land revenue.

(*iii*)   The farmers are free to avail the defence of innocent infringement.

(iv) As and when the breeder grants a sub licence for production of seeds, the farming community entitled to benefit sharing and can also have say in granting of sub licenses.

Chapter VII lays down elaborate provisions for the grant of compulsory licenses under the Act. Section 47-53 deals with compulsory licensing; and is, by and large, similar to the procedure under Patents Act.

Chapter VIII and IX talk about other procedural matters such as the Appelate Tribunal under the Act, finance, account and audit (Section 54-63).

Chapter X lays down elaborate rules relating to infringement and comprises section 64-77. The Act provides for both civil and penal provisions punishment for offences specified under the chapter may extend to Rs. 5 lakhs by way of time or imprisonment upto two years. In case of selling varieties with false denomination, there can be a minimum mandatory 6 months sentence.

Chapter XI comprising Sections 78-97 lays down miscellaneous provisions which may not be of any consequence to a non-lawyer.

GLOSSARY

'A'-line : The male sterile parent used in a cross to produce hybrid seed and also male sterile line.

'B'-line : Fertile counter part of the 'A' line which is called as maintainer line.

Barrier isolation : Growing tall plants like *Sesbania*, Dhaincha etc. all around the seed production plot to control the out crossing.

Basic Seed : Seed produced through mass selection (with progeny test) in a pure line variety or clone; source of breeder seed.

Breeder Seed : Seed produced by the originating or sponsored breeder of SAU's or ICAR Institutes. It is a source of foundation seed. Its quality is monitored by a monitoring team. The colour of tag is golden yellow.

Certified seed : It is produced from foundation seed. Its purity is certified by State Seed Certification Agency and is usually used for commercial crop production. The colour of the tag is Azure blue.

Composite Varieties : Advance generation of multiple crosses between selected varieties of diverse origin. Here the base population is used in compositing will be heterogeneous.

Cytoplasmic-genetic male sterility : A single dominant gene for fertility can over come the effect of cytoplasm where it is responsible for sterility.

Cytoplasmic male sterility : This is entirely controlled by the action of cytoplasm. Since the cytoplasm is transmitted through the female gamet only, this sterility is transmitted only through the female parent.

Detasselling : Removal of the tassel (male inflorescence) from the female parent in hybrid seed production of maize before its shedding pollens.

Double Cross : It is a first generation seed resulting from a cross between two single crosses. *e.g.,* (A × B) × (C × D).

Double Top Cross : First generation seed resulting from a cross between a single cross and an open pollinated variety *e.g.,* (A × B) × OPV.

Essentially Derived Variety : A variety which is predominantly derived from another variety and conforms to the initial variety in all aspects except for the differences which result from the act of derivation and yet is clearly distinguishable from such initial variety.

Extant Variety : A variety which is notified under Section 5 of the Seeds Act, 1966 or a variety about which there is common knowledge; or any other variety which is in the public domain.

Farmer : Any person who cultivates crops by cultivating land himself or by directly supervise the cultivation of land through any other person.

Farmers Variety : A variety which has been traditionally cultivated and evolved by the farmers in their fields; or is a wild relative or land race of a variety about which the farmers possess the common knowledge.

Foundation Seed : Seed produced from Breeder seed. Its quality is certified by State Seed Certification Agency. It is a source of certified seed. The colour of the tag is white.

Genetic Male Sterility : This is conditioned by a single recessive gene (ms). Dominant gene Ms result in the production of fertile pollen. The recessive male sterile gene is introduced into a line which is to be used as female parent.

Inbred : A nearly, homozygous line developed by continued inbreeding, usually selfing, accompanied by selection.

Isolation Distance : Distance to be maintained between the seed crop and the contaminants for the production of pure seed.

Maintainer Line : A line which is used for maintaining cms line. It has the same nuclear genotype as the male sterile line.

Multiple Cross : It is combination of more than four inbred lines. They are of high adaptability.

Nucleus Seed : A small quantity of seed with the breeder for the production of breeder seed.

Off-type : A plant or seed which deviates from the original characteristics as described by the breeder.

Out Cross : It is an Off-types plant resulting from a parent of a different sort contaminating the seed field.

R-line : Line which causes the resulting hybrid to be male fertile.

Rogue : An Off-type plant which is undesirable.

Roguing : Removal of Off-type plants from the seed plot.

Row Ratio : The ratio of number of rows of male parent to that of the female parent in the hybrid seed production.

Seed : The ripened ovule consisting of embryo and coats with often some additional food.

Single Cross : Cross between two inbred lines. The F_1 is high yielding and quite uniform *e.g.*, A × B.

Staggered Sowing : To synchronize the heading date of the male and female parents, especially for the hybrid combinations having parents with quite different growth duration. In order to extend the pollen supply time, the male parent is usually seeded twice or thrice at an interval of 3-4 days.

Supplementary Pollination : To enhance the extent of out crossing by shaking the pollen parent so that the pollen is shed and effectively dispersed over the 'A' line plants.

Synthetics : Advance generation of multiple crosses between selected inbred lines.

Tassel : The staminate inflorescence of maize.

Three-way Cross : It is a first generation seed of a cross between a single cross and an inbred line. *e.g.,* (A × B) × C.

Topcross : It is a first generation seed resulting from a cross between an inbred line and an open pollinated variety *e.g.,* A × OPV.

Variety : An assemblage of cultivated individuals which are characterized by growth, plant, fruit, seed or other characters by which it can be differentiated from other individual plants of the same kind.

Volunteer Plants : Plants emerged in the seed plot from previous crop.

INDEX

www.ingramcontent.com/pod-product-compliance
Lightning Source LLC
Chambersburg PA
CBHW072249210326
41458CB00073B/917